场 所 景 观
成 玉 宁 景 园 作 品 选

PLACE LANDSCAPE ARCHITECTURE
CHENG YUNING LANDSCAPE ARCHITECTURE WORKS

成玉宁·著

中国建筑工业出版社

图书在版编目（CIP）数据

场所景观：成玉宁景园作品选 / 成玉宁著. -- 北京：中国建筑工业出版社，2015.10

ISBN 978-7-112-18541-2

Ⅰ. ①场… Ⅱ. ①成… Ⅲ. ①景观设计－园林设计－作品集－中国－现代 Ⅳ. ① TU986.2

中国版本图书馆CIP数据核字（2015）第229676号

本书包括风景环境、城市公园、滨水景观、公共景观、住区景观、景观建筑六大部分共 24 个案例的介绍。主要论述了作者对于场所景观的理解，"走向场所景观"既是历练也是策略。景园设计与场所之间存在耦合关联，设计应从环境中生发而来。对景园环境既有秩序的研究不仅是对场所精神的响应，还是景园特色生成的源泉，更是场所本身之于景园设计的意义所在。基于耦合的景园设计方法不仅有助于寻求景观特征的生成逻辑，还是一种认知环境与设计景园的途径。

责任编辑：陈　桦　杨　琪
责任校对：张　颖　关　健

场所景观—成玉宁景园作品选

*

中国建筑工业出版社出版、发行（北京西郊百万庄）
各地新华书店、建筑书店经销
北京顺诚彩色印刷有限公司制版
北京顺诚彩色印刷有限公司印刷

*

开本：889×1194毫米　1/12　印张：34⅓　字数：600千字
2015 年 10 月第一版　2015 年 10 月第一次印刷
定价：318.00 元
ISBN 978－7－112－18541－2
　　　（27773）

版权所有　翻印必究

如有印装质量问题，可寄本社退换

（邮政编码 100037）

成玉宁

1962 年 4 月出生于南京市，1984 年获南京林学院园林专业农学士学位，1984 年 8 月至 1987 年 8 月，就职于南京市园林规划设计研究所，1990 年、1993 年分别获东南大学建筑系工学硕士、博士学位，1993 年起于东南大学建筑学院任教，现任东南大学建筑学院教授、博士生导师、东南大学建筑学院景观学系主任、东南大学景观规划设计研究所所长、东南大学风景园林学科带头人、东南大学建筑学院学术委员会副主任。

兼任国务院学位委员会第七届学科评议组成员、国务院学位委员会教育部人保部风景园林专业学位研究生教育指导委员会委员、全国高等学校土建学科风景园林专业指导委员会委员、中国风景园林学会理事、中国风景园林学会理论与历史专业委员会副主任委员、中国风景园林学会教育工作委员会副主任、江苏省土木学会风景园林专业委员会副主任委员、《中国园林》杂志编委。

成玉宁从事风景园林规划设计、景观建筑设计、景园历史及理论、数字景观及技术等领域的研究 30 余年，主持多项国家及省部级科研项目；著有《现代景观设计理论与方法》《湿地公园设计》等，主编《园林建筑设计》《中国园林史教程》等教材以及《数字景观》《现代风景园林理论与实践》等丛书；发表学术论文 60 余篇；设计作品获国家及省部级优秀设计奖 20 余次。

Cheng Yuning

In April, 1962, Cheng Yuning was born in Nanjing. After he got his bachelor degree of Agriculture from Nanjing Forestry University in 1984, he worked in Nanjing Landscape Architecture Planning and Design Institute for 3 years. Then he obtained his master degree and Ph.D. for Engineering in 1990 and 1993 respectively. In the same year when he hot his Ph.D., he started his teaching career in School of Architecture, Southeast University. And now, as a professor, Ph.D. supervisor and dean of Landscape Architecture Department, he has become an academic leader and vice chairman of the School's Academic Committee.

Professor Cheng has several academic appointments, including committee member of the Landscape Architecture Discipline Appraisal Group, Academic Degree Committee of the State Council; committee member of the Postgraduate Educational Instruction Committee of Landscape Architecture, Chinese Ministry of Education and Academic Degree Committee of the State Council; committee member of the of National Specialty Committee of Landscape Architecture, Civil Engineering Disciplines; council member of the Chinese Society of Landscape Architecture; vice chairman of the Specialized Committee of Theory and History, Chinese Society of Landscape Architecture; vice chairman of the Education Committee, Chinese Society of Landscape Architecture; vice chairman of the Specialized Committee of Landscape Architecture, Jiangsu Society of Civil Engineering; editorial board member of "Chinese Landscape Architecture".

Professor Cheng has dedicated to the study of landscape planning and design, architecture design in landscape, landscape history and theory, digital landscape and technology for more than 30 years; presided over a number of scientific research projects at national, provincial and ministerial level. He has written many famous books, including "The Theory and Method of Modern Landscape Design", "Wetland Park Design", etc.. And he is also the chief editor of several collections like "Landscape Architectural Design", "The History of Chinese Landscape Architecture", "Digital Landscape" and "The Theory and Practice of Modern Landscape Design". By now, he has published over 60 scientific papers, and his design works have won national, ministerial and provincial excellent design awards for more than 20 times.

序

中国园林艺术源远流长，始于商周，经历代耕缀，遂成蔚然大观，成为世界三大风景园林体系之一。其以独特、优秀的民族风格立于世界园林之列，为世人所称道。回溯中国园林的发展历程，传承与发展是为根本。一方面传统是相传成统，是从无到有、持续发展的。另一方面，传统又是随着时间不断积淀而成，因此需要不同时代的创新加以补充发展。

改革开放30年来，西风强劲，一时间非"西洋"即"南洋"，国人有时倒偏重于学习外国现代化先进技术的经验，而对如何结合学习中国传统的特色加以发展则重视不够。外国的经验要学，学他们如何按自然环境和历史人文的特色结合现代社会生活而创作。不要误认为现代的必然是欧美的。中国具有独特优秀的民族传统，在风景园林方面积累了丰富的经验。这与汲取外国的先进经验并不矛盾，古今中外皆可为我用。以中为体，以洋为用，风景园林设计也当充分挖掘中国园林艺术的经验，妥善处理现代化与民族、地域的矛盾，结合时代的特征发扬光大中国园林的精髓，实现"中而新"的创作目标。

在传承中创新从而满足现代社会生活的需要，创新势在必然。传承与创新是辩证的统一，创新与随意（恣意妄为）有根本的区别，即前者是"有根有据"，后者则"自圆其说"。创新是园林艺术的生命，创新的本质是适应新时代人类社会生活的发展。冶园不易，需要风景园林师与环境、场所的潜心对话。中华民族几千年来形成和不断完善的宇宙观和文化总纲是"天人合一"，指自然与人的统一。风景园林设计师总任务是保护自然环境和补充人造自然环境的不足。因此，"人与天调，天人共荣"的理念对风景园林师至关重要。

欧阳修说："钱塘莫美于西湖；金陵莫美于后湖。"风景园林是一市一地"美"的代表。风景园林是要将社会美融于自然美，从而创造风景园林的艺术美。"一方风水养一方人"，城市、建筑、园林创作的源泉在于"环境"，要求风景园林师理解环境，从环境出发，以本土环境融入各地方的人文来彰显中国特色和地方风格。景以境生，因境成景。因地、因境赋形特别重要。精在体验用地之异宜，布置自然和人文，融汇成具象的景和意境。"臆绝灵奇"的高级设计境界不是一蹴而就的。将设计工作化整为零，而每一环节都孜孜以求，呕心进取。风景园林需要密切结合时代的需要，将生态环境效益、社会效益和生产效益融为一体，并融入文化内涵。以设计为载体，保护自然环境和人，并结合人工再造自然环境以满足城市人民日益提高的物质和精神方面对生存环境的综合诉求，为人民长远和根本的利益服务，为民生的社会福利服务。

承前启后、与时俱进是当代中国风景园林师的历史天职。作为新时代的风景园林师，必须着眼两端：一端是学习、研究、继承优秀的民族文化传统，另一端是把握时代需要和精神。成玉宁君坚持"有思考的设计与基于设计的思考"，坚持基于风景园林学的自律性，30年潜心对话场所，持之以恒逐渐形成了系统的景园观与方法论，并且付诸实践，在工程中检验并升华为设计理论体系，自成一家。实践必以正确的理念为指引，成功的实践要提升为理论，再以理论指导新的实践。从场所出发正是"天人合一"的宇宙观在景园观中的体现，成玉宁君所选择的正确道路就是"承前启后，与时俱进"。因此，我祝贺《场所景观——成玉宁景园作品选》付梓出版，其中集结了他30余年潜心研究与实践的成果；更肯定其坚持的基于场所的景园设计观念与方法。基于本土实践与思考，从实践中来到实践中去，实现了从"实践——理论——再实践"的良性循环。望玉宁君巩固既有成绩，克服前路困难，在探索中国特色的新园林之路上取得更多成果，共尽风景园林设计师神圣的天职！

孟兆桢

中国工程院院士
中国风景园林学会名誉理事长
北京林业大学教授、博士生导师
2014年11月26日

PROLOGUE

Derived from the Shang Dynasty (1600~1046 B.C.) and Zhou Dynasty (1046~221 B.C.), the long-standing Chinese classical garden art has developed through different dynasties, and gradually stepped into its glorious phase, becoming one of the three landscape architecture systems in the world. It stands out among the world gardens for its unique and brilliant national style, which fascinates people all over the world. The retrospect to the evolution of Chinese garden indicates that inheritance and development are the foundation. On the one hand, tradition is formulated by inheritance, from nothing to a sustainable development; on the other hand, tradition is accumulated with the elapse of time, therefore its development needs innovations from different times.

In the past 3 decades since the reform and opening-up in China, there has been a strong trend of "Westernization". Chinese people preferred learning modern and advanced technology from western countries to inheriting and developing traditional features. It's necessary to learn from other countries, and what is the most important is to learn how westerners combine the modern social life with the environmental and historic features in landscape design. It is a misunderstanding that modern beauty can only be created by the west. China has its unique cultural tradition, and has accumulated rich experience in landscape architecture. Inheriting our tradition has no contradiction with learning from western culture. Taking their technique for practice, while basing Chinese culture as the core, landscape design should fully excavate the experience of Chinese classical garden art, properly handle the contradictions between modernization, region and nationality, carry forward the essence of Chinese classical garden in line with the contemporary characteristics, so as to achieve the aim of both "Chinese and Modern".

Innovation is an inevitable trend, while inheritance is aimed at meeting the needs of contemporary social life. Innovation and inheritance are a dialectical unity. Innovation is completely different from profligacy in that the former has a deep roots, while the latter only makes out a good case by itself. It is apparent that landscape design is not an easy thing, which requires the designers to have a deep "dialogue" with environment and places. "Unity of Human and Nature" is a general culture principle and world view of Chinese. Protecting natural environment and making up for the deficiency of artificial environment are the general tasks for landscape architects, therefore the concept of "man in harmony with the nature, sharing prosperity with the nature" is of crucial importance for landscape architecture designers.

Ouyang Xiu, a very important writer in Chinese literature history, wrote that "the most gorgeous scenery of Hangzhou is the West Lake, of Nanjing is the Xuanwu Lake". Landscape represents the image of the city and the region. Landscape should integrate social beauty into natural beauty, and then create the artistic beauty of itself. Human beings are shaped by the land they live in. The sources of creations in city, architecture and landscape are all based on environment. Landscape architects should be able to understand the environment, thus to integrate local environment with diversified humanistic implications to highlight Chinese characteristics and regional styles in every design work. Scenery is based on environment, to generate form based on site and environment is of great importance. The essence of design is to comprehensively understand the difference of sites, to precisely arrange both nature and cultural elements, to create artistic scenery and conceptual place. The design competence of "ingenious conception and smart layout" is not accomplished at one stroke, but developed from careful consideration given to each pieces which make up a complete design work. Taking design as a carrier, landscape architects should take the eco-environment benefit, social benefit and production effectiveness and culture connotation as a whole, and meet people's material and spiritual demands for their long-term and fundamental interests.

The obligation of contemporary Chinese landscape architects lies in not only linking the past and the future, but also keeping pace with the times. They need to deeply study the excellent Chinese cultural traditions, while precisely grasping the needs and spirits of the time. Mr. Cheng Yuning, who has persisted in conducting deep "dialogue" with places for 3 decades, insists on the idea of "design with thinking and thinking based on design", and puts emphasis on the self-discipline of landscape design, thus gradually developing his own viewpoints and methods of landscape design, which have been proved in practice. Practice need to be guided by proper theory, and successful practice should be concluded into theories that can be used to guide further practice. Based on local practice and thinking, he created a virtuous cycle for making theory and the actual union. The approach taken by Cheng is "linking the past and the future, keeping pace with the times", and his concept "design based on places" just embodies the general principle of "Unity of Human and Nature" of Chinese world. Therefore, I would like to express sincere congratulations for the publication of "Place Landscape Architecture: CHENG YUNING Landscape Architecture Works", as a result of his 30 years' painstaking research and practice. I hope that Mr. Cheng would reinforce his accomplishments and make further explorations in Chinese landscape development, to fulfill the sacred bounden duty for all contemporary Chinese landscape architects!

The Chinese Academy of Engineering
Honorary President of the institute of "Chinese Landscape Architecture"
Beijing Forestry University, Professor, Doctoral Supervisor
26/11/2014

CONTENT　　目录

TOWARDS LANDSCAPE OF PLACE　　8　　走向场所景观

SCENIC AREA LANDSCAPE　　风景环境

DA SHI LAKE ECO-TOURISM RESORT PLANNING, NANJING	22	南京市大石湖生态旅游度假区规划设计
LANDSCAPE ARCHITECTURE DESIGN OF SAKURA PARK, NANJING	36	南京市钟山风景名胜区樱花园设计
DESIGN OF DINGXIANG GARDEN IN YUHUATAI SCENIC SPOT, NANJING	58	南京市雨花台风景名胜区丁香园设计
LANDSCAPE AND ARCHITECTURE DESIGN OF NORTHERN NIUSHOUSHAN SCENIC SPOT, NANJING	68	南京市牛首山风景区（北部景区）景观及建筑设计
SHAJIABANG WETLAND ARCHITECTURE AND LANDSCAPE DESIGN, CHANGSHU	96	常熟市沙家浜东扩湿地景观与建筑设计

URBAN PARK　　城市公园

CONCEPTUAL PLANNING AND DESIGN OF SHITOUCHENG HERITAGE PARK, NANJING	116	南京市石头城遗址公园概念性规划设计
WUYI LAKE OFFICE PARK DESIGN, DAFENG	130	大丰市五一湖商务公园设计
GINKGO LAKE PARK DESIGN, DAFENG	144	大丰市银杏湖公园设计
ZHONGZHOU PARK DESIGN, HUAI'AN	160	淮安市中洲公园设计
YANGTZE RIVERSIDE LANDSCAPE DESIGN, YANGZHONG	174	扬中市滨江公园设计

WATERFRONT LANDSCAPE　　滨水景观

RIVERSIDE PARK DESIGN, SUQIAN	198	宿迁市河滨公园设计
ERMAOYOU RIVERSIDE LANDSCAPE DESIGN, DAFENG	210	大丰市二卯酉河景观带设计

PUBLIC LANDSCAPE　　公共景观

PUBLIC SQUARE DESIGN, TONGLU	226	桐庐广场设计
JINLING GRAND CANAL INTERNATIONAL CONFERENCE CENTER AND HOTEL LANDSCAPE DESIGN, SUQIAN	238	宿迁市运河金陵国际会议中心景观设计
JINLING RIVERSIDE HOTEL LANDSCAPE DESIGN, NANJING	252	南京市金陵江滨酒店景观设计
XINJIEKOU ZHENGHONG SQUARE LANDSCAPE REDESIGN, NANJING	274	南京市新街口正洪广场景观环境改造设计

RESIDENTIAL LANDSCAPE   住区景观

NEW WORLD GARDEN LANDSCAPE DESIGN, NANJING	290	南京市新世界花园景观设计
LANDSCAPE DESIGN OF HUACAI JIAYUAN IN WUYISHAN CITY	304	武夷山市华彩家园景观设计

LANDSCAPE & ARCHITECTURE   景观建筑

ARCHITECTURE AND LANDSCAPE DESIGN OF LAOSHAN FOREST PARK, NANJING	316	南京市老山森林公园景观及建筑设计
METRO LINE 1 GROUND STATION ARCHITECTURE AND LANDSCAPE DESIGN, NANJING	336	南京市地铁一号线地面站站厅及景观设计
ZIQING LAKE CROCODILE EXHIBITION HALL ARCHITECTURE AND LANDSCAPE DESIGN, NANJING	356	南京市紫清湖鳄鱼馆景观建筑设计
SU-WAN BORDER REGION GOVERNMENT SITE CONSERVATION PLANNING AND DESIGN	374	苏皖边区政府旧址保护规划与设计
WUQIU FARMING CULTURE CENTER OF GULI TOWN, CHANGSHU	386	常熟市古里镇坞丘农耕文化馆
LONGGUANG PAVILION DESIGN, HUAI'AN	398	淮安市龙光阁设计

POSTSCRIPT   408   后记

走向场所景观

成玉宁

引言

场所形成于特定的时空条件之中,既蕴含着自然的印记,亦记录着人类活动的信息。生成于场所中的景观是复杂的自然过程和人类活动在大地上的烙印,因此,几乎每一处场所均具有可识别性,显现出景园环境内在的独特性。画家凭借自我的艺术素养与技巧在空白的画布上挥毫泼墨,这一过程几乎不受时空和场地的限制。景园设计不存在"绝对的白纸",无论是自然环境抑或是历史文脉,皆是景园设计的依据和资源,也在一定程度上引导并约定了景园的创作。纯粹的艺术特色往往是艺术家对个人风格的追求,众多建筑师、景园师亦是如此:强烈的个人风格已然成为了许多大师的"标签",体现出独特且富有时代感的设计逻辑和审美倾向。景观环境的特色不宜仅仅指向设计师的个人风格或某种思潮,也不能简单地凭空创造,而应当有"根"有"据":回归到场所本身去寻求景园设计的源泉与创作灵感,彰显出特定的民族和地域特征,并揭示出场所中所蕴含的自然过程及人文脉络。

设计在一定程度上反映了时代及设计师本人的人居环境价值观。一方面,对环境进行改造的最根本目的在于满足人的需求,即所谓的"以人为本"。人在景园场所内的行为模式和使用方式都有一定的规律可循,是"人的秩序"。另一方面,景园设计不仅要满足人的使用和审美诉求,更应当符合自然材料本身的"存在规律"。景观环境的多重秩序对应着景园设计的多目标性,也是现代景园设计的特点之一。由此,景园设计是通过人为干预,在既有秩序基础上形成"新场所秩序",以满足多目标设计需求的过程。景观环境可持续性的实现在很大程度上有赖于场所内"原生"秩序与"新生"秩序之间的和谐共生与协调发展。客观上说,景园设计在一定的时空条件下拆解、重组多元的环境要素,人为改变地形、植物、石材等自然素材的位置、形状及组合方式,来赋予场所特定的"意义"。除了娴熟地运用形式和空间法则之外,景园设计在运用有生命的素材时要更多地兼顾其自然属性,尊重自然规律。20世纪初开始的"城市美化"运动,其目的在于创造一种新的物质空间环境,营造和谐之美。但因其偏重于形式而忽略了大众的需求,被认为是特权阶级为自己在真空中做规划,一味追求装饰性和纪念性却未解决城市的根本问题,也忽略了客观的生态效应,犹如"昙花一现",很快退出了历史舞台。作为阶段性的产物,城市美化运动不可避免地具有历史局限性,正是对其的反思促进了现代景园价值观的形成与发展:尊重场所的既有秩序,整合生态、空间、功能、文化等各个层面以满足多元化的需求。与之相应,当代景园设计的目的在于建构多元和谐共生的有机系统。

1. 场所的景观意义

不同学科对于"场所"内涵的理解不尽相同。诺伯舒兹(Christian Norberg-Schulz)从建筑学的角度提出场所的三个基本内容:其一,功能和结构;其二,对人的适应;其三,独特性和特殊性。就风景园林学而论,场所有着更为丰富的意义。景园设计是一个研究场所条件、寻求问题解决途径的过程。场所(Place)并非场地(Site),与代表纯粹物质空间的场地不同,场所具有物质与精神双重意义。场所是景园价值的重要载体,涵盖生态、空间、功能、文化四个方面的内容。任何一处场所都有其独特的内涵,即所谓的"场所精神"。场所精神铸就了环境的可识别性,决定了景园环境的个性与特征。

1.1 生态意义

场所中的构成要素及其状态是客观存在。其中,有生命的构成要素是景观环境的重要组成部分,受制于客观条件下的自然演替规律。因此,遵循客观规律是景园设计中不可忽视的前提。生态系统由生物群落及其环境构成,生物多样性和要素关系的复杂性,决定了景观生态环境是一个由多要素和多变量构成的复杂层级系统。自然环境中的动植物在适应气候、土壤以及地貌等条件的过程中,逐渐进化并与环境建立起协调的关系,生成了自身特定的演化规律与阶段性的稳定形态,具有自我修复与更新能力。对于景观环境的认识,需要有"阶段"意识,必须建立动态的观念,既要满足当下,更要适应未来演替的趋势。

"生态"之于景园设计涵盖了科学技术、价值观念以及审美取向等诸多方面,大大超越了原初的内涵,亦超出了"研究生物与环境之间相互关系"的范畴,而是包含了环境保护、可持续、生态化等多种层面的意义,从而变得更加广义和丰富。景园生态不仅仅是科学,更不止于技术,而是一种规划设计的"智慧",旨在景园环境系统的全生命周期内,通过最少的人工干预获得系统效益的最大化。

1.2 空间意义

空间是景观的载体,功能在其中展开,文化从其中生成,生态系统由其承载。景园空间不同于建筑空间,它没有完整、固定的"外表皮"。景观的空间界面是变化的，空间则是"多孔的"。对于景园空间问题的探讨，重点在于研究空间的内部，而非"外表皮"。与建筑空间相比，景园空间更加关注、强调贯通和交互，具有明显的拓扑效应；景园空间又不同于纯自然空间，它介于人工与自然之间，具有不确定性及复杂性，向内有其自身的构成关系，向外则与环境寻求关联。"内"与"外"的共同作用确定了景园空间的整体结构关系。生态、功能、文化都需要在同一空间中实现，人的活动、生境的演替以及文化的表达均通过空间展开。

1.3 功能意义

景园设计目的之一在于创造人性化的空间环境，满足不同人群的行为需求。人在场所中的行为是场所和人交互作用的结果。行为是多种因素交织在一起的综合产物，环境与人的行为之间存在着一定的对应关联，良好的公共空间可以促进人们的交往，丰富人们的户外生活；特定的空间形式也会吸引特殊的活动人群，诱发相应的行为与活动。景园设计师应充分研究场所中人的行为规律以及心理特征，依据不同行为的共性与规律展开"适应性"设计。行为与心理变化取决于人的需要和对周围环境的反应。人在场所中的行为既受"场力"的影响，同时又反作用于场所，与空间环境密切联系、相互制约。设计归根结底是为人服务的，因而具有强烈的人本意识。景园空间的设计需要通过人为的手段，调整干预环境构成要素的比例、秩序、形态等，从而合乎生境、功能与审美的诉求，综合地实现为人服务的目标。

1.4 文化意义

场所中的文化实质上是一种人对于土地的观念。场所景观强调的"场所认同"是指人与土地之间无法割舍的、在生存上和精神上的联系。重视场所的文化意义，其最终目的是延续环境中独具特征的人地关系，也就是所谓的"场所文脉"。由于自然及人为的作用，场所在历史中形成了印迹，具有唯一性与特殊性。场所之间的差异是导致景观多样性的内在因素。费孝通先生说，文化是"乡土的（From the Soil）"。景观环境作为一个绝佳的窗口，是建成环境中最具文化表现力的载体。无论是古典园林一类的历史文化遗产，抑或是反映区域地理特征的自然景观，均记录着长期以来自然或人工对场所的塑造过程。景园设计旨在甄选出有意义、有存在与传承价值的场所文化，通过保护、延续、重组到新的场所中。甄选的关键取决于文化本身的价值与意义，"今天"的设计应当从"昨天"提取符合时代审美的构成要素并加以强化。当代社会的意识形态、社会结构、生产生活方式、科学技术手段以及大众的审美需求同样是地域文化的重要组成部分，需要在历史传承和积淀的基础上，实现场所文化的更新和丰富，使新陈文化在景园中融合与共生。

2. 场所的景观特征

景园环境是复杂、多元的，采取的设计策略应有所差异。面对着自然风景与建成环境两类不同的对象，如何生成各具特色的景园环境？归纳看来，中小尺度的建成环境可以供设计师更多地发挥才情。劳伦斯·哈普林（Lawrence Halprin）借鉴东方的智慧——"搜尽奇峰作谱记"，大量地使用清水混凝土来表达山水趣味；丹·凯利（Dan Kiley）热衷于运用周匝建筑语汇，通过延伸建筑的柱网到外部空间中生成一体化的景园环境；彼得·沃克（Peter Walker）擅长利用简洁、现代的形式表现古典元素，实现平面艺术与景园设计的完美地结合；彼得·拉兹（Peter Latz）醉心于戏剧化地表现工业遗存……凡此种种均是设计师强烈的个人风格在景园设计中的体现。对于风景环境而论，情况却大相径庭，环境的"主宰者"由人转变为自然，凸显特征的一条普适性原则在于解读场所——发掘场所特质与传承场所精神。场所在漫长的历史过程中积淀了大量的特质化信息，这些独一无二的特征有助于营建特色化的景园。

2.1 遗传学的启示

风景园林是科学与艺术有机统一的学科。景园设计作品与纯艺术品不同，受到自然规律的制约。景观环境与生物体一样，有着自身的发展规律和内在的逻辑关系，因此，景园设计可以从探究生物体发展规律的遗传学中获得启示。

2.1.1 特征的必然性

现今地球上有约60亿的人口，基因的不同造就了个体间的差异，每个人都有唯一的面孔、指纹等识别特征，使得人类社会能够有序地运行。同样，场所也具有独一无二的"识别码"，它们是铸就景园环境基本特征的依据。

2.1.2 特征的遗传性

生物体性状特征的生成受控于基因，遗传信息的传递使后代出现与亲代相似的性状。景观环境中亦存在着类似的"遗传因子"——自然的土地、气候、生态、植被等各方面因素以及文化的印记。这些要素如同基因一般，共同组成影响景观环境特征的内在因素。生物个体对异源的器官具有排异反应，基因差异越大，排异反应越明显。同理，异源的景观要素可能与原场所之间产生矛盾和冲突，对景观环境的品质与可持续产生影响。只有依据内在的逻辑和客观规律生成的景园特色才能够与环境和谐、持续共生。

关于基因结构、功能及基因表达的研究始终是遗传学关注的焦点。基因的突变与重组是生物变异发生的根本原因，也是形成生物体多样性的重要原因之一。场所固有性状的传承可使景观环境承续其原有的特征，而对要素进行变形和重组，则可以生成更为多元化的景观特征。现代科学研究表明，遗传基因的改变是不定向的，只有通过科学的调控与筛选才能实现生物品种的改良。同样，景观环境中特征的延续并非是对场所特征的盲目复制，而是需要在满足客观规律和新系统秩序的基础上，进行合理的优化与更新。通过解读，将场所特征转译成景观的"基因编码"，并选择具有价值的部分加以呈现，是形成景园环境可识别性与特色的有效途径。

2.2 特征的构成

2.2.1 生态学特征

自然环境要素是场所固有的组成部分：山地、高原、丘陵、盆地、平原等地貌类型对景园空间具有重要的影响；光照、降水、风等气候条件对景观风貌的形成有着巨大作用……自然环境要素决定了场所的生态学特征。热带雨林、热带疏林、沙漠、温带草原、温带森林、寒温带针叶林等植被气候带中植被种类组成、群落结构、季相变化均大相径庭，各具特色，所产生的生态效应也各不相同。场所的生态学特征自古以来就被人们所认知和利用，人居环境的外在形式因所处自然环境的不同而变化万千：传统的北方园林与江南园林在景象上显示出鲜明的地域区别；黄土高原的窑洞与西南地区的吊脚楼也因自然环境和资源条件的截然不同而产生了形式上的巨大差异。

2.2.2 形态学特征

形态学是一个广义的概念，在生物学、艺术学、数学等看似毫不相干的学科中都占有一席之地。形态学方法立足于现象学原理和构形论思想，研究的是事物的表象特征和内在的构成原理。在风景园林学科中，形态学更多地应用于场所的空间形态及格局的研究，用以探索内在关联与规律。场所的空间形态是环境实体所表现出来的具体物质形态，往往是最直观、最易识别的要素。空间是多变、灵活、多义的，景园设计中空间形式的生成与原初的场所空间形态有着密切的关联。场所的空间形态深受自然环境的影响，自然要素作为风景环境的主体，往往占据了景园空间的主导地位。因而，风景环境的空间是动态变化的，季节的更替、植物的生长均会使原有的空间形态发生变化，如在夏季围合感强的内向型空间，冬季往往因为植物的凋谢而呈现对外的开放性与渗透性；又如枯水期可穿越的河道，在丰水期却成为不可逾越的空间界限。此外，场所的整体空间形态亦受到区域地貌的影响。场地周边的地形地貌特征不仅限制了场地的围合程度，对小气候也有一定的影响。空间形态与结构之间存在着一定关联。空间结构指景观要素在空间上排列和组合的形式，反映了空间形态构成的内在规律。对场所空间结构进行探讨，有助于把握场所空间的本质性特征，在设计中生成适宜的空间表达方式。

2.2.3 文化学特征

文化是人们长期创造而形成的产物，凝结在物质之中又游离于物质之外。场所是一个"容器"，承载着人类活动留下的信息，人文内涵是场所的重要标志。孔子的《论语•雍也》有云："智者乐水，仁者乐山。智者动，仁者静。知者乐，仁者寿。" 形象地表述了"山水"环境对于人性格的塑造。在历史长河中，人类按照自我意识对自然环境施加影响，由此赋予了场所独特的文化学特征，体现为民族性、地域性和时代性。东亚的中华文化是汉字文化圈的核心，起源于农耕文明，价值观念深受儒、释、道思想体系的影响，行为模式受以礼义道德和宗法制度的制约，对待自然讲求"天人合一"的态度。这种文化氛围衍生出独特的中国传统园林体系，以古典园林为代表，追求"师法自然"，融入了中国古代山水画的意境，讲求"一石以代峰，一池以代水"，园林空间的营造强调周而复始、循环往复以实现无穷的境界，体现了时人的生活观和环境观。起源于西亚的伊斯兰园林体系，深受伊斯兰教文化的影响，展现出了截然不同的价值取向和自然观念：以天堂为范本，在现实沙漠中塑造"绿洲"，在空间形式上多运用十字形的林荫路和水渠构成中轴线，将封闭的建筑与特殊的水系相结合，并以精美细密的图案和装饰色彩作为建筑特征。当代景园设计需要以客观全面、动态发展的眼光来看待场所的文化学特征，避免一味地强调物质遗存而忽略精神内涵，也不可墨守陈规而拒绝新文化的植入，追求文化的可持续性和多元融合应当成为景园设计的主流价值取向。

2.3 景园特征的生成依据

2.3.1 逻辑依据

逻辑思维是人脑对客观事物间接概括的反映，凭借科学的抽象来揭示事物的本质，具有自觉性、过程性、间接性和必然性的特点。在逻辑思维中，一切事物都存在系统结构，所有信息均以系统的方式加以组织。场所作为一个系统，包含了诸如生态、空间、功能、文化等子系统，子系统内部不仅有着树状的纵向层级关系，而且相互之间也存在着千丝万缕的横向交叉与关联。归纳与演绎、抽象与概括、分析与综合、组合与分解等是景园设计的基本逻辑思维方法。自然界是一个不可分割的整体，场所是整体的一个部分，与大环境紧密关联。因此，保证场所与整体环境的和谐、遵循客观规律是前提。同时，实现场所个性也是设计的重要目标，景园特征从场所中生成，但绝非场所特征的完整复制。该过程类似于人类的繁衍，父母的基因遗传至下一代，子女与父母在诸多方面有着相似性，但绝非是对父方或母方个体的克隆。

2.3.2 形象依据

"形象"一词总是与感受、体验关联在一起。景观特征的生成过程中，形象思维与逻辑思维并非是矛盾的对立面，而是互为补充与依托。依据形象思维生成的景观特色具有生动性、直观性和整体性的特点。景园设计基于对场所客观形象体系的感受，结合设计师的主观认知和个人情感进行识别（包括审美判断和科学判断等），用针对性的形式、方法描述景观形象。景园设计不应满足于已有形象的再现，而应致力于对场所既有形象的加工与升华，从而获得新的形象特征。这一加工过程具有跳跃性和创造性，反映出了景园设计的艺术性。东西方古典主义造园都注重模仿的价值，不论是摹写自然山水，抑或是具体的物象，甚至是再现故事情境，建构形象与意义之间关联是景园设计表达的主要途径。场所感是人对特定环境中某种经历的感受，基于对物象的情感认同，延续、更新和彰显场所中的原有形象，可强化场所感。

3. 场所的解读

倡导"走向场所景观"的目的在于通过对场所的解读，顺"势"而为：尊重生态演替规律、构建空间秩序、满足行为需求、传承场所文脉，通过适度的设计实现景园环境的整体优化，彰显与强化场所的固有特征，从而保证景观的多样性，丰富人居环境。

3.1 解读场所

景园设计中的"因地制宜"具有其特殊的意义。"宜"建立在对环境的理性认知基础之上，因而对场所固有"信息"的解读便成为景园设计的基本前提。

3.1.1 生境解读

场所生境的解读应综合考虑两个方面，一是自然环境要素，包括地质、地貌、气候、水文、土壤、植被、动物等方面；二是人类活动及其影响，包括土地利用方式等。原生环境具有自身的稳定性与可持续性，对于生境质量良好、生态系统稳定的自然区域，需要在遵循自然规律和环境内在机制的基础上，通过保护相对稳定的生态群落和空间形态来维护生态系统的演化能力。通过定性、定量、定位相结合进行分析与评价，确定场所中不同地块的生态敏感性与建设适宜性，划分保护、利用、优化的区域，从而采取相对应的规划设计策略。在保护生态环境的基础上有选择地利用自然资源，将人为过程有机地融入自然过程之中，使"因地制宜"落到实处。

3.1.2 空间解读

形态、结构、界面、肌理是构建景园空间的四个基本要素。空间形态与结构生成了场所基本形式。景园空间反映了场所的变迁过程，体现景观环境的整体性特征。景园空间的界面呈现虚实相间的状态，因而空间具有一定的开放性。拆分和重组是丰富外部空间的基本手法。场所的肌理在自然和人为的双重作用下积淀而成，所以一方面具有自然的属性，客观地反映了自然过程，呈现出规律性；另一方面还饱含人文的精神，体现了人们对环境的使用要求与意志。场所空间整体性的建构，需要同时满足多种秩序的要求，以实现场所综合效应的最大化。同时还应加强与既有空间格局和形态的联系，创造出具有形式和意义的新空间，实现"源于自然而高于自然"的景园设计目标。

3.1.3 行为解读

人在景园环境中的行为有着特定的模式与规律。行为与环境存在着互动效应，环境可以诱发行为，行为亦会反作用于环境。对不同空间形态中人的行为加以观察，可归纳出人们在景园环境中一般性的行为规律，进而预判待建场地上人的活动方式与行为特点，从而解读场所中潜在行为的可能性。实现景观园环境的"人性化"应当从整体出发，基于人群的心理与行为活动特征与规律，调整景园构成要素以适应场所潜在的行为需求。

3.1.4 文脉解读

景园环境的"文脉"对应于人类活动积淀于场所之中的"印记"。无论是历史悠久的农业文明还是近现代的工业文明，都留下了极富历史价值的遗存；历经岁月洗礼的传统村落与见证了城镇发展的历史街区，皆反映出不同时期的传统风貌和特定地域条件下的人地关系。场所中的文脉既有具象的物质文化遗存，也有抽象的文化精神。"有形"的文脉内涵广泛，包括自然过程及土地使用模式，"无形"的文脉则是人们的集体记忆。需要尊重场所原有的自然过程、格局和人类活动留下的历史文化积淀，并以此为本底和背景，在保证历史脉络线索"完整性"的基础上，充分发掘景观环境的既存特性，甄别、筛选不同时期的人类使用模式、具有特殊意义的事件等，选择其中可发展、可利用的部分，通过引申、变形、嫁接、重组、抽象等手法，重新组织到新的景园秩序之中。在保护与利用场所遗留"印记"的同时，融入新的场所信息，使"新生"的景园空间具有"生于斯长于斯"的特质。

3.2 尺度的意义

"尺度"是一个较为宽泛的概念，不单纯以大小论，也不仅仅用来描述空间。人居环境科学范畴内，城乡规划学旨在解决城市、区域及以上尺度的问题，它为城市设计、街区设计乃至单体建筑设计明确大方向；建筑学旨在解决城市设计及其以下尺度的问题，即从建筑群体到单体的细部。风景园林则不同，几乎覆盖了从单体、节点直至规划层面的问题，因此风景园林设计是一种"全尺度"的设计。所谓"全尺度"，即并非单纯地研究某一层面，而是从不同层面研究人与生境、文脉在空间中的共生问题。

风景园林根据讨论对象综合特征的不同，可分作自然风景环境和人工建成环境两大类型。前者以自然为中心，侧重于探讨可持续利用自然资源的途径；后者以人为中心，致力于整合区域内的自然资源与人工环境，提升人居环境的整体效能。对于风景环境与建成环境的研究需要把握"尺度"差异性，从而发展出适宜的景园规划设计策略。

3.2.1 自然的尺度

风景环境更多地反映着自然的规律与进化历程，具有地域性、持续性等特征。自然尺度的风景特征不以人的意志为转移，其变化主要受制于自然要素与规律。遵循自然尺度原则的风景环境规划设计应服从于自然过程，采取"最少干预"的方法，尽量减少人为扰动造成的影响，并通过规划设计促进自然系统的物质利用和能量循环，维护原有的生态格局。

3.2.2 人的尺度

建成环境中的景园空间与人的关系最为密切，以人工营造为主体，以服务于人为宗旨，因此遵循"人的尺度"原则。如同普罗泰戈拉（Protagoras）所言"人是万物的尺度"，东西方景园设计均带有鲜明的人本意识。设计师可从不同的角度出发，彰显景园的个性与风格，在不断突破传统中，创造出新的形式与内涵。除了追寻个性化和独创性之外，建成环境中的景园设计更要以人为本，满足人们的诉求，激发人们的归属感。

4 耦合的法则

对风景园林而言，所谓"设计"即最大限度地整合场所的资源，以最少的人为干预实现预期的设计目标，强调在设计中最大程度地利用环境资源和自然"力"。与之相对应，强调"减量化"设计是在集约化的规划设计思维体系架构下，基于合理的人为干预，达成对景观环境资源的优化配置。"耦合法"从场所出发，贯穿于调查、分析、设计与建造的全过程，通过对各个环节的有效调控达到减量化的设计目标，对当代景园设计具有重要的现实意义。

4.1 "因地制宜"释义

人们喜用"因地制宜"来描述对环境认知与利用的方法。"因地"强调的是设计从场所出发，针对场所展开设计；"制宜"指的是在设计中进行适当的选择与优化。"因地制宜"包含了两个基本面：一方面指场所利用的最大化，即尽可能利用场所中的既有资源，发挥场所的最大效益；另一方面是对场所扰动的最小化，在完成设计目标的同时最低程度地对场所进行干预，反对过度设计与营造。"因地制宜"的设计可以实现对场所的集约化利用，体现了可持续的景园设计理念。

4.2 耦合法的原理

物理学中"耦合"指的是两个或两个以上的系统或运动方式通过各种相互作用而彼此影响以至联合的现象，是在各子系统间的良性互动下，相互依赖、相互协调、相互促进的动态关联。"耦合"的概念包含了系统、关系和动态三个方面。现代景园规划设计基于学科的本体呈现系统化、多目标的特征，因而必须统筹生态、空间、文化、功能四个基本面以形成有机统一的整体。将"耦合"的基本理念引申到景园设计中来，就是强调对场所的尊重及自然力的运用，在多设计目标与场所固有的秩序和要素之间形成关联，利用环境资源并提升环境整体的品质。"耦合"不是对场所属性的改变，而是最大限度地弥合"异源性"的元素与"本源性"的场所。耦合法覆盖了从本体到形式、功能到技术、设计到建造，直至后期管养的全过程。

不同环境、不同层面、不同尺度实现"耦合"的手段及所采取的适宜技术各异。在大中尺度环境下，耦合法作为一种方法论，与"地域主义"相比更具有可操作性和系统性。小尺度中，耦合法提倡"与环境的对话"，这与"有机建筑"理论存在相似性。但"有机建筑"意在寻求建筑与环境之间的整体和谐，更多关注的是建筑自身的问题，具有单向性。而耦合法体现了动态的互适性，强调生态、空间、功能、文化与场所之间的对应关联。"耦合"关注的不止于形式上的和谐，而是新植入元素与环境的"无缝衔接"。"耦合"是一个动态的过程，在此过程中设计目标与场所互相影响，最终达到和谐共生。因此，设计目标与场所之间的互适性是耦合的原则。"互适"即互相适宜、适应。这里的"适宜性"是双向的，一个层面指设计目标积极主动地与场所的适应，即根据环境选择适当的设计项目与设计手法等；另一个层面是对环境进行适度改造。

耦合法的意义主要包括了三个方面：一是最大程度地体现场所的固有特征，有助于传承场所特质，是设计实现个性化、特色化的基本渠道和策略；二是通过耦合可以实现设计的减量，以最小化干预来满足场所要素的重组需求；三是科学技术手段辅助下的场所认知更为客观、准确，有利于将景园设计由基于感觉引向知觉。

4.3 基于耦合法的规划设计策略

以系统论为指导，耦合法致力于构建具有自我完善能力以及自律性结构的整体。对景园设计而言，"耦合"就是根据环境条件选择适宜的项目，再将其引入到场所中去，与之对应的是当下规划设计中的策划环节；第二个环节对应于总体规划，即形成规划总图，依据不同项目选择适宜的环境，区分地块、沟通路网、优化竖向；第三个环节是对单一项目的调控，包括景观建筑的规模、覆盖率、体量、高度、建设强度等控制要素，即规划设计中的控制性详规层面。耦合法的三个环节分别对应了风景园林规划设计中"项目策划—总体规化—修建性详规"阶段，层层递进地覆盖了整个风景园林的规划设计过程。

4.3.1 项目的场所适宜性原则

景观项目的确立属前期研究，在考虑市场诉求的同时还需要结合景园设计的基本理念，基于对于现场的研究确立景观项目的内容和定位。前期研究要求具有前瞻性，对可能发生的游憩内容加以预判。景观项目应从场所中寻求项目存在的依据，即研究项目与环境的关联性。风景环境中项目的设立在考虑上位规划、政府决策等相关导向性要求的同时，还应充分考虑、游憩的因素，其中包括市场需求、游人量、目标人群等方面。除此之外，场所中蕴含的文化资源也是项目内容确立的重要依据。景观项目不是"无本之木"，更非"无源之水"，"生长"于场所的项目更具合理性与可持续性。将适宜的项目依据功能和定位加以分类，结合场所调研后的定性、定量分析，可确定规划设计的分区，实现项目和特定区域间的耦合。

4.3.2 设计要素与场所的耦合

耦合的第二个阶段是在场所要素分析、评价的基础上将各设计要素与场所进行耦合的过程，对应于规划分区建立后的进一步深化工作，也是规划总图的初步形成过程。要素与场所的耦合是将上一过程中确立的具体项目建立起与特定场所的对应关系，即"选址"。设计需要厘清例如道路、建筑、水体等主要目标要素与场所的耦合关系。无论是道路选线还是建筑选址，抑或水景的营造，均与坡度、坡向、汇水等场所条件息息相关，因此对场所的认知是探讨设计要素与场所之间耦合关系的第一步。对场所的研究要从空间、生态、文化等层面出发，细分至高程、坡度、植被类型等最为基本的因子层面，通过GIS等辅助分析手段与方法，充分认知场地，完成生态敏感性、建设适宜性等场所评价研究。最终因地制宜地确立场所与项目的对位关系。

4.3.3 设计层面的调控

"耦合"是一个全尺度的概念，涵盖了大尺度中项目选择、落地直至小尺度范围的建筑和景观设计。如何控制建构筑物的尺度与体量，一直是景园规划设计的难点。经过要素选址后，设计的视野聚焦于较小的尺度，需要从与环境对话的角度出发，对景观以及建筑进行控制，包括了建筑色彩、建筑风格、建筑体量、建筑高度、建设强度、建筑群体空间组合形式、建筑轮廓线、景观视线等方面。从环境出发的设计调控能够更好地将建筑与人工景观"缝合"到自然环境中。不仅如此，在场所分析与评价基础上的控制性耦合还对环境容量进行综合研判，从空间容量、生态容量等角度对人为改造活动进行考量，在不突破环境承载力的前提下，发挥自然力的最大效益。

4.4 基于耦合法的多方案优选

几乎所有的设计都是"多解"的，设计师对同一场地的认知、主观愿望以及设计能力的差异，都会影响设计过程及设计方案的生成与发展。长期以来，人们习惯将不同的设计方案加以比选，对应于时下设计市场上流行的方案招标，目的就是在不同方案间寻求更与场所特质契合的佳作。离开了场所评价的方案比选往往难逃"选美"的命运，失去了比选的初衷。耦合的法则有助于在不同的方案中选择出与场所最为契合的构思。设计方案之于场所的耦合度是评价方案优劣的基础。

5. 走向场所景观

"走向场所景观"既是理念也是策略。在场所解读的基础上形成适宜的设计思路，其意义远远不止于空间及形式特色的形成，更在于对特定区域生境条件的响应、潜在行为的满足、特色文脉的传承……这是源自景园设计自身特性的选择。景园设计与场所之间存在耦合关联，设计应从环境中生发而来。对景园环境既有秩序的研究不仅是对场所精神的响应，还是景园特色生成的源泉，更是场所本身之于景园设计的意义所在。基于耦合的景园设计方法不仅有助于寻求景观特征的生成逻辑，还是一种认知环境与设计景园的途径。

TOWARDS LANDSCAPE OF PLACE

Cheng Yuning

Introduction

Place is formed in a specific time and space, bearing a natural imprint and recording human activity. Landscape in a place is the mark of complex natural processes and human activities on the earth; therefore, almost every place is recognizable, showing the inherent uniqueness of garden environment. Unlike well-trained painters who can brush on a blank canvas freely, almost unlimited by time and space, landscape architects must design a garden according to its natural environment and historical context, to a certain degree, these resources guide and define garden design. Purely artistic specialty is often the artist's pursuit of personal style, so do many architects and landscape architects. A strong personal style has become the "trademark" of many masters, reflecting the uniqueness in contemporary design logic and aesthetic tendencies. However, the characteristics of garden environment should not be directed only to the designer's personal style or some ideas, cannot be simply created out of nowhere, but should be well-based. The latter refers to seeking the inspiration of garden design from its place, highlighting the features of nationality and region, and revealing the natural traits and human context in a place. "Dialogue" and "symbiosis" are the trend of modern garden design.

Design, to a certain extent, reflects the habitat environment values of the times and the designers. On the one hand, the ultimate goal to transform the environment is to meet the needs of people, the so-called "people-oriented." There are rules for people to follow in a garden concerning their behavior patterns and the methods of use, the so-called "human order." On the other hand, garden design does not only meet people's practical and aesthetic appeal, but shall comply with the law of natural material. Multiple order of landscape environment corresponds to the multi-objective in garden design, one of the characteristics of modern garden design. Thus, by human intervention, garden design is the process to form "new place order" in an established order, to meet the needs of multi-objective design. The sustainability of garden environment depends largely on the harmonious symbiosis and the coordinated development between the "native" order and "new" one in a place. Objectively speaking, garden design is, under certain conditions of time and space, dismantling and restructuring diverse environmental elements, by using terrain, plants, stone and other natural materials to change their position, shape and composition, to give a place a specific " meaning. " In addition to skillful use of form and space, garden design needs to pay special attention to the use of materials with life and their natural attributes, respecting the laws of nature. In the early 20th century, "city beautification" campaign aiming to create a new physical space environment, and achieve a harmonious beauty, but because of the emphasis on form while ignoring the needs of the public, is considered to be the planning of the privileged class out of touch with the reality. The blind pursuit of decorative and commemorative style failed to solve the fundamental problems of cities, also ignored the objective of ecological effects. Thus like "flash in the pan", the campaign was soon withdrawn from the stage of history. As a stage product, city beautification movement inevitably has a historical limitations, and it is the reflection on the movement that promotes the formation and development of modern landscape values: to respect the established order of places, to integrate ecology, space, function and culture at all levels in order to meet diverse needs. Correspondingly, the purpose of the contemporary garden design is to establish a pluralistic and harmonious organic system.

1. Landscape significance of a place

For the word "place", different disciplines have different interpretations. From the architectural point of view, Christian Norberg-Schulz proposed the three basic elements of a place: First, the function and structure; second, human adaptation; third, the uniqueness and particularity. As far as landscape architecture is concerned, place has a richer meaning. Garden design is a process to study the conditions of a place and to seek ways of solving problems. Place is not a site, not a purely physical space. In fact, a place entails both the material and spiritual meaning. Place is an important carrier of landscape value, covering the contents of ecology, space, function, and culture. Every place has its unique content, the so-called "spirit of place." Place spirit creates the environment recognizability, and determines the personality and characteristics of garden environment.

1.1 Ecological Significance

Elements and their status are the objective reality in a place. Among them, elements of life are an important part of landscape environment, subject to the natural laws of succession under objective conditions. Therefore, observing the natural law is the premise in garden design. Ecosystem is composed of biological communities and their habitat, and the overall property is not a simple sum of the elements, but the interaction between them. The biological diversity and complexity determine that the ecological system of landscape be a complex hierarchy system of a multi-element and multi-variable. Flora and fauna in the natural environment to adapt to climate, soil and topography and other conditions in the process, gradually evolve and establish a coordinated relationship with the environment, developing their own law of evolution and stable form of a particular stage, to form a self-repair and upgrading ecosystem. To understand the landscape environment, the notion of "stage" and dynamic concept should be taken into consideration, to meet the immediate need as well as to adapt to the future trends in succession.

As far as garden design is concerned, "ecology" covers science and technology, values and aesthetic orientation and other aspects, far beyond the original meaning, even beyond the scope of "the relationship between the biological and the environment". It refers to the meaning of sustainability, ecological balance, environmental protection and other aspects, and thus its meaning is broader and richer. Garden ecology is not just science and technology, but a "wisdom" of planning and design, aiming at the whole life cycle of garden environmental systems, to maximize system efficiency with minimal human intervention.

1.2 Space significance

Space is the carrier of landscape, from which garden's functions are unfolded, culture is developed, and ecosystem is evolved. Unlike the architectural space, garden space does not have a complete, fixed "outer skin", lack of a clearly define boundary. The space interface of landscape is changing, and its garden space is "porous." The discussion concerning garden space focuses more on the study of its internal space, rather than the "outer skin." Compared with the architectural space, more attention and emphasis is given to the through and interactive space, with obvious topological effect; garden space is different from the purely natural space, and found between artificial and natural space, with uncertainty and compound characteristics, having relationship inwardly with its own and outwardly with its environment. The joint action of "inside" and "outside" determines the structural relationship of garden space. Ecology, function, and culture all need the same space to be materialized, and human activity, habitat succession and cultural expression all need space to stage on.

1.3 Functional significance

One of landscape design goals is to create space environment of humanity to meet the behavioral needs of different groups. The human behavior in a place is the result of the interaction between place and human. The human behavior is an intertwined comprehensive product under the influence of a variety of factors. There is an association between certain environment and human behavior. Good public space can contribute to people contacts, enrich people's outdoor life; specific spatial form will attract special group of people, induce the corresponding behaviors and activities. Garden designers should fully study the law of behavior and human psychological characteristics in a place, conduct the "adaptive" design according to the common features of different behaviors. Behavioral and psychological changes depend on the needs and reactions of people to the surrounding environment. Human behavior in a place is influenced by the "field force", but also is counterproductive to the place. Human behavior is closely connected with the space environment and mutual restraint with each other. In a nutshell, design is for people, hence should be people-oriented in nature. The design of garden space is conducted by people, through adjusting and intervening the proportion, order, shape and other aspects of environment elements to meet the habitat, practical and aesthetic appeal, to realize comprehensively the goal to benefit people.

1.4 Cultural significance

Culture in a place is essentially the concept of people about their land. Landscape in a place emphasizes on "place identity", referring to the inseparable connection between people and land in terms of survival and spirit. The ultimate goal to attach the importance to the cultural significance of places is to continue in their habitat the unique feature of the relationship between people and land, also known as the "place context." Due to natural and man-made factors, places leave unique and special imprints in history. Differences between places are the intrinsic factors leading to landscape diversity. Mr. Fei Xiaotong (A Chinese sociologist) said that culture is from the soil. Landscape environment as a great stage is the most culturally expressive carrier in a built environment. Whether it is historical and cultural heritage like classical garden, or natural landscape reflecting the regional characteristics, they have recorded either natural or artificial shaping process of places for a long time. Garden design aims to select the place cultures with meaningful and inheritance value, by protecting, extending, reintegrating them into the new places. The key of selection depends on the value and significance of culture itself. The design of "today" should extract the sensible aesthetic elements from "yesterday" and strengthen them. Ideology, social structure, way of production and life, science and technology of contemporary society, as well as the aesthetic needs of the general public are also an important part of regional culture, and it is necessary to renew and enrich culture of places on the basis of historical heritage and accumulation so that the old and the new culture can be integrated and coexist in the garden.

2. Landscape features of places

The complexity and plurality of garden environment create different design strategies, facing with two different types of objects, namely, the natural landscape and the built environment. How to create distinctive garden environment? The answer depends on the type of environment. Generally, it seems that a small scale built environment can bring talented designers to much fuller play. Lawrence Halprin drawing from Eastern wisdom - "search for the best spectral peaks in mind," used heavily as-cast-finish concrete to express his landscape taste; Dan Kiley was keen to use cloister, by extending building column grid to the outer space environment, to create the integration in a garden; Peter Walker specialized in the use of simple, modern form to express classical elements to achieve the perfect combination of graphic art and garden design; Peter Latz obsessed with the dramatized performance of the industrial heritage ...... All these in the designers are strong personal style reflected in garden design. For landscape environment, the situation is very different, for the dominator of environment is changed from people into nature, which is obviously not the people's will. A universal principle highlighting the characteristics is to interpret places – to explore the characteristics and the spirit of places. The accumulation of a large amount of particular information in places in the long course of history help construct the characteristics of landscape.

2.1 Inspiration of genetics

Landscape is the organic unity of science and art. Garden design works are different from those artistic ones, subject to the laws of nature. Like organisms, landscape has its own law of development and internal logic; therefore, garden design can draw inspiration from genetics that explores the law of the development of organism.

2.1.1 Inevitability of characteristics

Today, there are approximately 6 billion people on the planet, and the differences in gene lead to the differences in individual. Each of us has a unique face, fingerprint and other identifying characteristics, thus human society can be orderly operated. Similarly, places also have a unique "identification code", and these codes are the bases for making the environment of places unique.

2.1.2 Hereditary of characteristics

Characteristics of organism are controlled by genes, passing genetic information to the next generation, bearing the same biological traits with their parents. There is also a similar "genetic factor" in landscape environment, as a constituent part of the environment, subject to land, climate, ecology, vegetation and other factors, carrying cultural imprint as well. These elements like genes in general, together constitute the internal factors controlling environmental characteristics of the landscape. Individual organism rejects allogeneic organ, the greater the genetic differences, the more obvious the rejection. Similarly, heterologous implantation landscape elements may cause contradictions and conflicts with those in the native place, thus influence the quality of the landscape environment and its sustainability. Garden characteristics that are created only on the basis of internal logic and objective laws can achieve the harmony with the environment, and sustainable symbiosis.

The study on gene structure, function and gene expression is always the focus of genetics. Recombinant and mutation of gene is the root cause of biological mutation, and it is one of the important reasons for the formation of biological diversity. The inheritance of the intrinsic elements of places makes landscape environment keep its original features, but the deformation and reorganization of their elements can create more diverse landscape features. Modern scientific research shows that genetic changes are not directed only to achieve improved biological species through scientific regulation and filtering. Similarly, the continuation of the landscape environment features is not done through the blind copy of the characteristics of a place, but to conduct reasonable optimization and upgrade, on the basis to meet the needs of objective law and the order of the new system. By interpreting, the characteristics of a place are translated into landscape "genetic code", and the valuable parts are selected and highlighted, which is an effective way for recognizability and characteristics in landscape environment.

2.2 Component of characteristics

2.2.1 Ecological characteristics

The physical environmental elements are a natural part of places: mountains, plateaus, hills, basins, plains and other landforms have a significant impact on the garden space; light, precipitation, wind and other weather conditions have a huge effect on landscape formation...... The physical environmental elements determine the ecological characteristics of places. Tropical rain forest, tropical woodland, desert, temperate grasslands, temperate forests, boreal forest and other vegetation and climate zones of vegetation species composition, community structure, seasonal variations are very different, and the ecological effects are also various. Since the ancient times, the ecological features of places have been perceived and utilized by people, and the external form of habitat environment varies due to the differences of the natural environment: there is a sharp distinction between the traditional northern gardens and the southern ones; Cave dwellings of Loess Plateau and the Hanging house the southwestern region of China are cases in point showing the impact of the differences on natural environment and resources.

2.2.2 Morphological characteristics

Morphology is a broad concept, found in seemingly unrelated disciplines such as biology, art, mathematics. Morphological methods is based on the principles of phenomenology and construal theory, studying the appearance and internal constitution of things. In landscape architecture, morphology is often used in the study of spatial shape and pattern in places, to explore the internal association and laws. Space form of places is the specific material form in environmental entity, often the most intuitive and most recognizable features. Space is varied, flexible, and polysemous, with different forms in different places. Space form in garden design is closely related with the original form of places. Space form of landscape places is greatly influenced by the natural environment, with the natural elements as the main body of the landscape environment, often dominating garden space. Thus, the space of landscape environment is dynamic, changing the original spatial form through the succession of seasons and the growth of plants. For instance, in summer, there is a strong sense of inward enclosure space, while in winter there is outward openness and permeability due to withered plants; and in dry season the river can be crossed, in rainy season it becomes insurmountable. In addition, the overall space shape of places is also affected by regional geomorphological effects and limitations. Regional landform is a complex system of multi-factor and multi-level, thus the landforms of places not only limit the degree of enclosure, also have some impact on the microclimate. There is some correlation between the shape and structure of space. Spatial structure refers to the form of permutations and combinations of landscape elements in space, reflecting the inherent laws of space form composition. The exploration of spatial structure of places helps grasp the essential characteristics of space, and in turn contributes to the appropriate spatial expression in design.

2.2.3 Cultural characteristics

Culture is the product of people's creation activity over history and existed in and beyond material world. Cultural connotation is an important symbol of places. Place is a "container", carrying messages left by human activity. Confucius once said in Analects:" The wise find pleasure in water; the virtuous find pleasure in hills. The wise are active; the virtuous are tranquil. The wise are joyful; the virtuous are long-lived. " He vividly expressed the" landscape "environment shaping the human character. In the long history of human, people have imposed their influence in on the natural environment based on their own will, which gives the place unique cultural characteristics, reflecting the ethnic, regional and epical features. The Chinese culture in East Asia is the core of Chinese culture and civilization, originated in the agricultural civilization, with values affected by Confucianism, Buddhism and Taoism, and people's behavior patterns by propriety and patriarchal system of moral constraints, adopting an attitude to nature by emphasizing the harmony between people and heaven. This culture gives birth to a unique system of traditional Chinese garden, with classical gardens as a representative, in the pursuit of "learning from nature", incorporating the ancient Chinese landscape painting into garden design, emphasizing "a stone in lieu of peak, in lieu of a pool of water." In terms of garden space, image of endless cycling of life and infinite extension of space, reflecting the view of people on life and Environment. The Islamic system of garden is originated in the West Asia, heavily influenced by the Islamic culture, showing different values and natural concept: with paradise as pink print, to create "oasis" in the desert in reality, in the form of space, cross-shaped tree-lined roads or canals form the axis, combining the enclosed building with a special water system, using fine design and exquisite decorative colors as the architectural features. Contemporary garden design needs to look at the cultural characteristics of places with an objective and comprehensive and dynamic development point of view, to avoid blindly emphasize the material remains while neglecting the spiritual content, and do not stay in a rut and refuse implantation of a new culture, the pursuit of sustainable and blending of culture should become the mainstream values of garden design.

2.3 Garden features foundation

2.3.1 Logical foundation

Logical thinking is the indirect generalization to objective things by human brain, by virtue of scientific abstraction to reveal the nature of things, with features of consciousness, process, indirectness and inevitability. In logical thinking, everything is systemic architecture, all the information is organized in a systematic way. Place as a system, includes subsystems such as ecology, space, function, culture, not only has a longitudinal internal subsystems hierarchy tree relationship, but also countless transverse cross associations with each other. Induction and deduction, abstraction and generalization, analysis and synthesis, composition and decomposition are the basic logic thinking methods in garden design. Nature is an indivisible whole, with places as parts of the whole, closely associated with the environment. Therefore, to achieve harmony between place and the overall environment, observing the objective law is a prerequisite. At the same time, it is an important goal to achieve place personality. Garden features are created from places, but by no means a complete copy of the features of places. The process is similar to human reproduction, passing genetic inheritance of parents to the next generation, thus children and parents resemble each other in many respects, but by no means are the father's or the mother's clones. Garden design must respect the objective laws.

2.3.2 Foundation of image

The term "image" is always associated with the experience and feelings. In the process of creating of landscape features, imagery thinking and logical thinking are not the opposites of a contradiction, but complementary and supportive to each other. Landscape features created in imagery thinking are vivid, intuitive and holistic. Imagery thinking should not be satisfied with the existing image reproduction, and should be committed to place both image processing and sublimation, to obtain new image features. The whole process is full of jump thinking and creativity, reflecting the artistry of garden design. The imagery thinking in garden design is to create and describe landscape image with targeted form and approaches, based on the perception and storage of the objective image system delivered by places, combined with the designer's subjective cognition and personal feelings identified (including aesthetic judgments and scientific judgments, etc.). Both the Eastern and Western classical gardens are focused on imitation value, whether depict natural landscapes, or specific objects, or even the reproduction of narrative context. The construction of correlation between image and meaning is the main way of expression in garden design. Sense of a place is a human experience and feeling in a particular environment, and established in a space of sense of identity. The continuity, upgrade and highlight of the original image in a place can enhance people's sense of identity to a place.

3 Interpreting places

The aim to promote the view of approaching landscape of places is to well fit places by interpreting them: respect the law of life succession, construct spatial order, meet the people's needs, pass on cultural context, achieving the overall optimization of garden environment through appropriate design, highlighting and strengthening the intrinsic properties of places, so as to ensure the diversity of landscape, and enrich people's living environment.

3.1 Interpretation of places

That landscape design should be the "adaptation to local conditions" has a special meaning, in which "adaptation" is found on the basis of rational cognition of the environment, and therefore the interpretation of the inherent properties of "information" of places has become the basic premise of garden design.

3.1.1 Habitat interpretation

Two aspects should be considered when interpreting the habitats of places: firstly, the natural environment elements, including geology, topography, climate, hydrology, soils, vegetation, animals and so on; secondly human activities and their impacts, including land use methods. Native environment features stability and sustainability. To the natural areas of good quality for habitat and stability in ecosystem, it is necessary to follow the laws of nature and the internal mechanism of the environment, to maintain the capacity of evolution in ecosystem through protecting the relatively stable ecological community and spatial form. It is necessary to conduct analysis and evaluation through qualitative, quantitative, and positioning methods to determine the suitability of ecology and construction of different plots in places, dividing, utilizing and optimizing the areas for protection in order to take corresponding planning strategy. On the basis of protecting the ecological environment, it is advisable to be of selective use of natural resources and integrate the human process organically into the natural process, and thus fully achieve the goal of "adaptation to local conditions".

3.1.2 Spatial interpretation

Shape, structure, interface, and texture are the four basic elements to build garden space. Spatial form and structure creates the basic forms of places. Garden space records the changing process of places, reflecting the overall characteristics of landscape environment. The form of space is derived from the enclosed interface. Garden space interface shows the status of intertwined solid and void, so the space possesses a certain openness. Mutual penetration, fragmentation and reassembly are the basic techniques to enrich outer space. Texture of a place is accumulated in the dual role of naturel and people, so on the one hand it has a natural property, objectively reflecting the natural process, showing regularity; on the other hand it is also full of the spirit of humanity, reflecting people's requirements and will on the environment. Construction of spatial integrity of a place needs to meet the requirements of a variety of orders to maximize the combined effect of the place. It should also strengthen links with the existing spatial pattern and form, to create a new space of form and meaning, and thus achieve the goal of "from nature and above nature" in garden design.

3.1.3 Behavior foreseen

Human behavior in the landscape environment has specific patterns and laws, and are interactive with each other. Environment may induce behavior, and behavior will be counterproductive to the environment. To observe people's behavior in a corresponding environment can generalize the behavior patterns of people in the landscape environment, and thus to predict the activity patterns and behavioral features of people in places to be constructed, so as to interpret the potential behavior of people in places. To realize the humanity of landscape environment, a holistic approach should be adopted, based on the characteristics and laws of people's psychology and behavior activities in landscape environment, adjusting garden elements to accommodate the potential behavioral needs of places.

3.1.4 Interpretation of cultural context

"Cultural context" of garden environment corresponds to the accumulation of "marks" left by human activities in places. Whether it is agricultural civilization with a long history, or modern industrial civilization, they have left us a cultural legacy with historical values; traditional villages and historic districts reflect the traditional style of different periods and human-nature relationship in a particular region. The cultural context in places includes both the physical and spiritual ones. The "tangible" cultural context is of broad connotation, including natural processes and land-use patterns, while the "invisible" context is the collective memory of people, needs to respect the original natural processes and patterns, and honor the historical and cultural accumulation left by human activities. With this background and context, it is necessary to ensure the "integrity" of historical development of context, to fully exploit the existing features of landscape environment, to screen the human use patterns in different periods, and the event with special significance, to select the parts that can be developed and used, to reorganize them into a new garden order, by way of extension, deformation, grafting, restructuring, abstract and other techniques. While protecting and exploiting marks left by places, the memory of places should be integrated with the information of a new place, so that the "new" garden space has a "born and bred" character.

3.2 Scale significance

"Scale" is a broader concept, neither simply about the size, nor confined to the description of space. In the scope of habitat environment, urban planning aims to solve the problems of urban, regional scale and above, providing a clear direction for urban design, street design and the design of individual buildings; architecture is designed to solve the problem of the urban scale and below, namely from groups of buildings to the details of individual buildings. Landscape is different, almost covering everything from individual, node till the problem of planning aspects; therefore, landscape design is a "full-scale" design. The so-called "full-scale", is not simply to consider the issue at one level, but to study the symbiotic problems in space between human, habitats, and context from different levels.

Depending on the comprehensive features of the subject for discussion, landscape can be grouped into two types, namely the natural landscape environment and artificial built environment. The former is centered on nature, focusing on the approaches to explore ways for the sustainable use of natural resources; the latter is people-centered, committed to the integration of natural resources and the man-made environment in a region, to improve the overall performance of habitat environment. For the research on landscape environment and the built environment is needed to grasp the differences in "scale", in order to develop an appropriate strategy of garden planning and design.

3.2.1 Nature of scale

Landscape environment reflects more on the law of nature and evolution, with regional, permanent and other characteristics, and the landscape features of natural scales are not changed to suit people's will, which is mainly subject to the major elements and the law of nature. Landscape environmental planning and design, following the principles of natural scale, should be obedient to the natural process, and adopt the method of "minimal intervention" to minimize the impact of human interference, and to promote the substance usability and energy cycling of natural systems through planning and design, maintaining the original ecological pattern.

3.2.2 Human scale

Garden space is the closest to human in a built environment, aiming to serve the human and with artificial buildings as the main body, so it follows the "human scale" principle. As Protagoras said, "Man is the measure of everything", people-oriented is clearly shown in both the Eastern and the Western garden design. From different point of view, designers continue to break tradition, highlighting the garden's personality and style, to create more new forms and meanings. In addition to the search for personalization and originality, it is more important to be people-oriented, to stimulate people's sense of belonging and sense of identity.

4 Coupling law

To landscape architecture, the so-called "design" is to maximize the integration of resources of places, with minimal human intervention to achieve the desired design goals, emphasizing in the design to maximize the use of environmental resources and natural "force." By contrast, emphasizing "reduction" in design is under the framework of integrated planning and design, on the basis of reasonable human intervention to achieve optimal allocation on the landscape of environmental resources. "Coupled Law" starting with places, throughout the investigation, analysis, design and construction of the whole process, through the effective control of all the aspects of design to achieve the reduction targets, has important practical significance to contemporary garden design.

4.1 Interpreting "adaptation to local conditions"

People like to use "adaptation to local conditions" to describe the method of environment cognition and utilization. Here, "cause" means the "basis", "land" means "place", "local conditions" emphasize that the design should reflect the features of its place; "adaptation to" refers to the proper selection and optimization of the design."Adaptation to local conditions " consists of two fundamentals: on the one hand it refers to the maximization of the use of premises, namely the fullest possible use of existing resources in a place, to achieve maximum efficiency of places; on the other hand it is to minimize the disturbance to places, in the completion of the design target, to minimize the intervene to places, to object to excessive design and construction. "Adaptation to local conditions"in design is designed to realize the integrated use of places, reflecting the sustainable concept in garden design.

4.2 Principle of coupling law

In physics, "coupled" refers to two or more systems or movements through various interactions with each other as well as the combined impact of the phenomenon, and under the benign interaction between the subsystems, is the dynamic association of interdependence, mutual coordination, mutual reinforcing. The concept of "coupling" includes systems, relation and dynamic. Modern garden design is based on the ontology feature of the discipline to become systematic planning and design, with features of multiple objectives, and thus must coordinate ecology, space, culture, and function to form an organic unity of the whole. The introduction of the basic ideas of "coupling" into garden design is to emphasize on the use of the natural forces and honor places, to form the correlation between the inherent elements and order of places and the multiple design goals, to use environmental resources to improve the overall environment quality. "Coupling" is not a change of place attributes, but to maximize the bridge between the "heterologous" element and the "endogenous" in places. Coupling method covers the whole process, from ontology to form, from function to technology, from design to construction, till the care and maintenance afterwards.

The approaches and technology to achieve "coupling" vary in different environments, levels and scales. In the medium and large-scale environment, coupling law as methodology, is more operational and systematic than "regionalism." In small scale, the coupling law promotes the "dialogue with the environment," which resembles the "organic architecture" theory. But "organic architecture" is intended to seek overall harmony between architecture and the environment, with more attention on the problems of buildings themselves, so it is one-way communication. The coupling law reflects the dynamic of mutual adaptability, emphasizes the correlation between ecology, space, function, culture and places. "Coupling" concerns not only the harmony of form, but the "seamless" cohesion between a newly implanted element and the environment. "Coupling" is a dynamic process, in which design goals and spaces affect each other, and ultimately to achieve harmonious coexistence. Therefore, the mutual adaptation between design objectives and places is the principle of coupled law. "Mutual fit" means suitable for each other, to adapt. Here the "suitability" is two-way, with one level referring to the design goal to proactively adapt to places, to choose the appropriate design projects according to the environment, proper design techniques and so on; the other level is the appropriate transformation of the environment.

The significance of coupling law mainly includes three aspects: Firstly, to maximize and pass on the inherent features of places in the project, is the fundamental channel and strategy to achieve personality and particularity in design; secondly, to achieve the reduction in design through coupling is to minimize the intervention in order to meet the needs of places recombinant; thirdly, the cognition of places assisted by the scientific and technological means is more objective, accurate, and helps the garden design change from feeling based activity to perception.

4.3 Planning and design strategies based on coupling law

Guided by the theory of system, coupling law is committed to construct the overall structure capable of self-improvement and self-discipline. For garden design, the term "coupling" is to select the appropriate project based on the environmental conditions, and then introduce it into the place, corresponding to the today's step of planning in garden planning and design; the second step corresponds to the overall plan, the formation of the total plan, depending on the various project to choose the appropriate environment, to distinguish block, to establish communication network, to optimize the vertical; and the third step is the control of a single project, including landscape architectural scale, coverage, volume, height, and other elements, namely the controlled detailed planning level in planning and design. The three steps of the coupling law correspond to the three stages, namely, "planning – overall planning – detailed planning" respectively, progressively covering the entire planning and design process landscape architecture.

4.3.1 Suitability principle of projects in places

The establishment of landscape projects is of the preliminary study, requiring people to take market as well as the basic concepts of garden design into account, to clearly define the content and position of landscape project, based on the study of places. Preliminary studies need to be prospective, to predict the possible recreational contents. Landscape projects should seek their foundation from places, namely the correlation between the project and the environment. The project of landscape environment is established in consideration of the relevant guiding requirements such as the master planning, the government decision-making as well as factors of tourism, including aspects of market demand, the number of visitors, and target groups. In addition, the cultural resources in places is also an important basis for the establishment of the project content. Landscape project is neither a tree "without root", nor a river "without source", projects "grown" out of places are more reasonable and sustainable. Suitable projects will be classified according to their function and position, in combination with qualitative and quantitative analysis of places after the, to determine the partition in planning and design, to achieve the coupling between planning and design projects and particular regions.

4.3.2 Design elements and coupling of places

The second stage of coupling is the process to couple the design elements into places on the basis of the analysis on places, which corresponds to the further deepening work after the partition in planning, namely the initial formation process of the overall plan. The coupling of elements and places is to establish the correspondence between a specific project established in the previous process with a particular place, namely "the selection of sites" The design should clarify the coupling relationship between the main target elements and places in terms of the selection of routes and construction site, as well as water-scapes and vertical optimization. Whether it is the selection of route and construction, or the creation of water-scapes, they are all closely related with slope, aspect, catchment and other conditions of places, so the first step to explore the coupling between design elements and places is to knwo the places. The study on places should be conducted from the macro-perspective of space, ecology, and culture, down to the micro-prospective of elevation, slope, vegetation type, etc., through GIS and other analytical tools and methods, to fully understand places, complete the evaluation of ecological sensitivity, and the suitability of construction sites and so on. Ultimately, the correspondence between places and projects are established with the principle of adapting to local conditions.

4.3.3 Control at design level

"Coupling" is the concept of a full-scale, covering the selection of a large scale project, project landing, and the architecture and landscape design of a small scale. For a long time in the landscape environment, it has been a problem in garden planning and design about how to control the landscape scale and the volume of buildings. After the selection of sites, the vision of design focuses on a smaller scale, requiring the dialogue with the environment, to control the landscape and building, such as color, style, volume, height, strength, the combinations of forms, contour, and landscape view. The design control, taking environment into consideration, is able to "stitch" buildings and artificial landscape into the natural environment. Moreover, the controlled coupling based on the analysis and evaluation of places also offers the comprehensive judgments on environmental capacity, from the viewpoint of space and ecological capacity, to ensure that human activities to the environment should not break its capacity, and thus bring the forces of nature into full play.

4.4 Selection of multi-projects based on coupling law

Almost all of the design are "multi-solution", and the differences in the cognition, subjective desire and ability to design of a designer to the same place will affect the formation and development of the design process and pattern. For a long time, people are used to compare and select between different designs, which corresponds to the popular bidding project in the design market nowadays, and its purpose is to seek best choice among different projects. The fate of this kind of comparison of projects is doomed to fail, without the consideration of places. Coupling law helps choose the best fit with places in different projects. The degree of coupling in places of the design is the basis for assessing the merits of the project.

5. Towards landscape of place

"Approaching landscape of place" is both a philosophy and a strategy. Appropriate thinking in design is formed on the basis of interpreting places, and its significance is far more than just the formation of features in space and form, but also in response to a specific area of habitat conditions, the inheritance of cultural context with special features…… which is derived from the selection of the characteristics in garden design. Garden is subordinate to the environment, and the design should come from environment. Research on landscape environment established order is not only a response to the spirit of places, but also a source of landscape features, and above all it is the place itself where the meaning of garden design is all about. It is an approach in environment cognition and garden design to seek the rationality where landscape features are created based on the coupling of garden design.

南京市大石湖生态旅游度假区规划设计
DASHI LAKE ECO-TOURISM RESORT PLANNING, NANJING

项目地点：江苏省南京市	Location: Nanjing, Jiangsu	项目成员
设计时间：2005年	Design Period: 2005	景观设计：成玉宁、陈 聪
建成时间：2006年	Completion Time: 2006	建筑设计：成玉宁、殷向明
委托单位：南京丰盛产业控股集团	Client: Nanjing Fengsheng Industry Holding Group	结构设计：薛 峰
用地面积：372公顷	Area: 372 ha	水电设计：刘 俊、刘向上

项目概况

大石湖生态旅游度假区位于南京市牛首山风景区北麓，总面积约3.72平方公里。景区内以丘陵地貌为主，南高北低，自然生境条件良好。原用地类型包括林地、苗圃、农田、裸地及建设用地，其中林地主要由人工针叶林、地带性次生阔叶林以及针阔混交林组成。度假区内部存在人工水库泄漏及农田常年水涝的情况，亟需改善。

设计策略

本设计坚持生态优先原则，在生态保护的前提下，优化配置场地资源，调整景观格局。在山水环境之间点缀休闲、健身、餐饮、娱乐等各类休憩设施，以满足人们亲近自然、调养生息、健康生活的需要。

设计的开展以场所适宜性评价为基础，坚持因地制宜，集约化利用土地资源。首先，采用数字化叠图对场所进行分类评价，分析各区域的生态适宜性，明确项目建设适宜性及用地范围，并采取相应的保护及优化措施。其次，依据地形地貌及汇水条件，梳理、整合过去因农业生产而改变的自然空间形态，延山引水以恢复场地的自然属性。设计利用自然高差，灵活营造出湖泊、湿地、溪流、跌水等形态各异、富于变化的水体景观，形成完整的园区水系，解决了区域内长期排水不畅的问题。其次，依据不同地段固有的空间特征，如山林、湖泊、溪流、池塘、坡地等，结合地带性特点配置适生植物。在优化生态环境的同时，构成了具有亚热带北缘地带性特征的生态群落景观。

Project Profile

Dashi Lake Eco-Tourism Resort is located at the northern foot of Nanjing Niushoushan Scenic Spot, covering an area of 3.72km^2. Dominated by hills where the southern section is higher than the northern, the resort is a favorable natural habitat. The original land use type included forest land, nursery, bare land and construction land, among which the forest land was made up of coniferous plantation, zonal secondary broad-leaved forest and theropencedrymion. However, in this resort, measures need to be taken to tackle problems, such as artificial reservoir leaking and the perennial waterlog of farmlands.

Design Strategy

Ecology is the priority of this design. With ecological protection as the precondition, land resources are optimized to achieve suitable landscape pattern. Dotted with various recreation facilities among the natural scenery for relaxation, body-building, catering and entertainment, the resort can meet people's demands of being close to the nature, having a good rest and living a healthy life.

Based on the suitability evaluation of the selected site, this design adjusts to local conditions in order to make proper use of the land resources. First, digital overlay map is used to give classified evaluation on the site and analyze the eco-suitability of different sections within the site to determine the construction suitability and the site scope, followed by corresponding protective and optimal measures. Second, changes of natural spatial form caused by agricultural production have been sorted out and categorized on the grounds of landforms and catchment conditions, and the site's natural characteristics are restored through extending mountains and channeling water. Taking advantages of the natural elevation difference, this design flexibly creates a variety of water landscapes like lake, wetland, stream and head fall, which make up a complete water system within the resort and solve the long-standing problem of poor drainage. Third, based on the fixed spatial characteristics of different sections, such as mountain, forest, lake, stream, pond and slope, plants are arranged in suitable living sections. Thus the design realizes both the improvement of the eco-environment and the formation of an eco-community landscape characterized by zonal features of the northern rim of the subtropical zone.

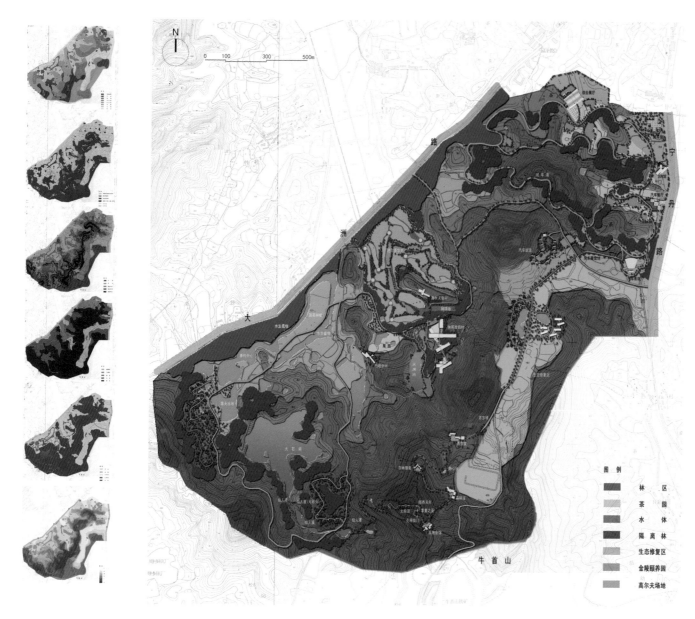

大石湖生态旅游度假区总平面图

1	3	1. 北部池塘鸟瞰
		2. 管理中心实景
2	4	3. 北部园景透视
		4. 桂花园漫水步道

1. 林间石径
2. 桂花园汀步
3. 林间憩台

风景环境　SCENIC AREA LANDSCAPE | 29

杉林湿地

湖景别墅实景

1. 树阵休闲广场
2. 湖畔会所实景
3. 湖景别墅实景

大石湖湿地群落

南京市钟山风景名胜区樱花园设计
LANDSCAPE ARCHITECTURE DESIGN OF SAKURA PARK, NANJING

项 目 地 点：江苏省南京市	Location: Nanjing, Jiangsu	项 目 成 员
设 计 时 间：2010 年	Design Period: 2010	景观设计：成玉宁、袁旸洋、赵 楠
建 成 时 间：2012 年	Completion Time: 2012	建筑设计：成玉宁、袁旸洋、鲍洁敏、侯汝凝
委 托 单 位：江苏省人大外事办	Client: Foreign Affairs Office of Jiangsu Provincial People's Congress	结构设计：盛春陵、周毅雷
用 地 面 积：37236 公顷	Area: 37236 ha	水电设计：王晓晨、李滨海

项目概况

樱花园位于江苏省南京市钟山风景名胜区，始建于 1996 年，原有面积 20 亩。园区内部原地形平坦缺乏变化，部分场地存在排水不畅问题，导致樱花长势不良；景观空间无序，原有的建筑及小品风貌杂乱。根据省政府要求，提升樱花园景观等级并扩大面积至 60 余亩，以樱花为意象表达"中日友好"的文化主题。

设计策略

1) 延山引水，优化场地空间格局

设计围绕"中日友好"的文化主题展开，采用"空间单元分置法"，将中、日不同风格的景观单元分别布置，通过组织游览线路、结合绿化的掩映与分隔，有意识地将景观单元相对片区化，又适度透景。同时，采取交互展开的空间单元与序列，巧妙、形象地展现了中日两国文化的源流关系。

2) 莳花植木，突出植物景观特征

樱花为全园的植物造景基调，园内栽植了 8 个品种的樱花 3000 余株；同时，将富有中国特色文化内涵的梅花置于同一园林空间之中；为丰富园区的季相观赏性，适当增植了不同花期的观花植物，补植地域性色叶、常绿植物。

3) 安亭置榭，彰显景观主题

设计通过布置具有传统江南园林建筑风格的六角攒尖重檐亭、歇山亭、石拱桥等展现中国的文化，亦设有体现日本传统园林特征的鸟居、洗手钵、石灯笼、赏樱亭等；另根据人物故事情节需要设置了具有民国风格的廊与花架，局部营造枯山水庭院。

4) 疏源掘流，传承场所文脉

设计从空间结构的把握入手，借助细节的表达和文化氛围的渲染，凸显场所特色和文脉。匠心独具地将文化线索"分而并置"："中国园"一侧以六尊明式螭首为引导，老井券与"日本园"一侧的洗手钵隐喻了中日文化"源"与"流"的关系；而具有民国建筑风格的友谊廊与紫藤花架，则隐喻着故事发生的年代和中日友好的主题。

5) 优化竖向，满足植栽及水景营造

园内水系原为人工硬质驳岸，由于处理不当导致水源难以存蓄；东部地段积涝致使樱花长势不良。因此，设计重点整理了园区水系，解决了水量存蓄问题，同时栽种水生植物，着力营造自然驳岸、跌水、漫滩的优美水景；对部分樱花栽植区域进行了竖向优化，从而满足樱花的生态习性。

Project Profile

Built in 1996, the Sakura Park is located in Nanjing Zhongshan Mountain National Park in Jiangsu Province, originally covering a total area of 20 mu (1 mu equals to 666.7m^2). The landform of this park used to be flat and lacked variety; some section even suffered from poor drainage which stunted the growth of sakura; and the landscape space was in a mess, with architectural structures and accessorial buildings showing no order. On the request of the provincial government, this project upgrades the park's landscape class and expands its area to over 60 mu, with sakura symbolizing the cultural theme of "China-Japan Friendship".

Design Strategy

1) Extend mountains and channel water to optimize the site's spatial pattern

Centering on the cultural theme of "China-Japan Friendship", the design utilizes "Separate Arrangement of Spatial unit", that is, to respectively arrange different styles of Chinese and Japanese landscape units through organizing tour routes and using green plants to set off or separate these units. This arrangement brings both proper indivisibility and relative separation to these Chinese and Japanese landscape units. Meanwhile, the design uses mutually unfolded spatial units and sequences to display the derivative relationship between Chinese and Japanese cultures in a clever and visual way.

2) Plant flowers and trees to highlight plant landscape characteristics

With sakura as the landscaping keynote, in this park, 8 species of sakura are planted, amounting to more than 3000 in total. Meanwhile, plum blossoms which carry unique Chinese cultural connotation are also planted. To improve the ornamental value in all seasons, flower ornamentals of different florescence as well as zonal color-leaved and evergreen plants are added.

3) Build pavilions to manifest landscape theme

The design manifests Chinese culture through traditional garden architectures usually built in regions south of the Yangtze River, such as hexagonal pavilion with pyramidal double eaves, pavilion with gable and hip roof and stone arch bridge. Besides, Japanese culture is also manifested through its traditional garden architectures like torii, finger bowl, stone lantern and sakura observation pavilion. In addition, corridors and pergolas are set with a combination of Japanese rock garden to correspond with characters and plots.

4) Date back to the origin to pass on the cultural context of the site

Having grasped the spatial structure, the design utilizes detailed expression and cultural atmosphere to highlight the site features and its cultural context. To show its originality, this design arranges cultural clues "separately but within one space". On Chinese Garden side, six Ming-style Chishou (hornless-dragon head) lead the way, while on the Japanese Garden side, the old well and the Japanese finger bowl create a metaphor for the relationship between Chinese culture, the "source", and Japanese culture, the "course". Furthermore, the Friendship Corridor and the Wisteria Pergola which are characterized by architectural style of the Republic of China (1912~1949) implies the year when this story took place and the core theme of China-Japan friendship.

5) Optimize the vertical space to meet the need of planting and creating water landscape

In this garden, the original water system was artificial hard revetment. Due to improper construction, it used to be difficult to reserve water, and frequent waterlog in the eastern section caused poor growth of sakura. Therefore, the design reorganizes the water system to solve the problem of water reservation. Meanwhile, hydrophytes are planted to create beautiful water landscapes of natural revetment, head fall and floodplain. Vertical optimization has been conducted to some sakura planting areas so as to cater for sakura's ecological habits.

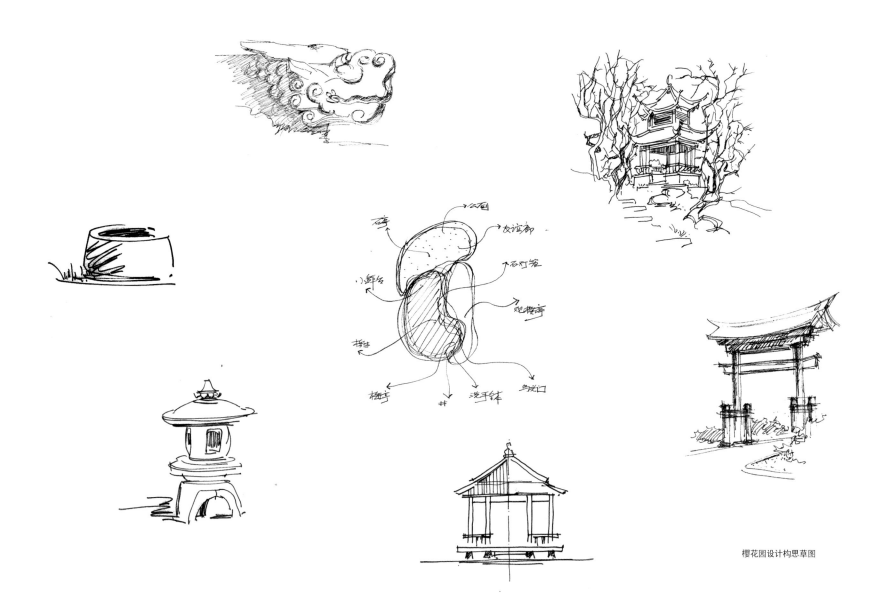

樱花园设计构思草图

樱花园总平面图

风景环境　SCENIC AREA LANDSCAPE | 39

友谊纪念碑及石灯笼实景

樱花园入口景观

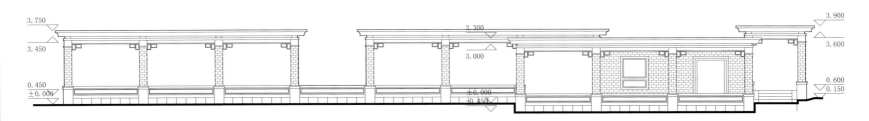

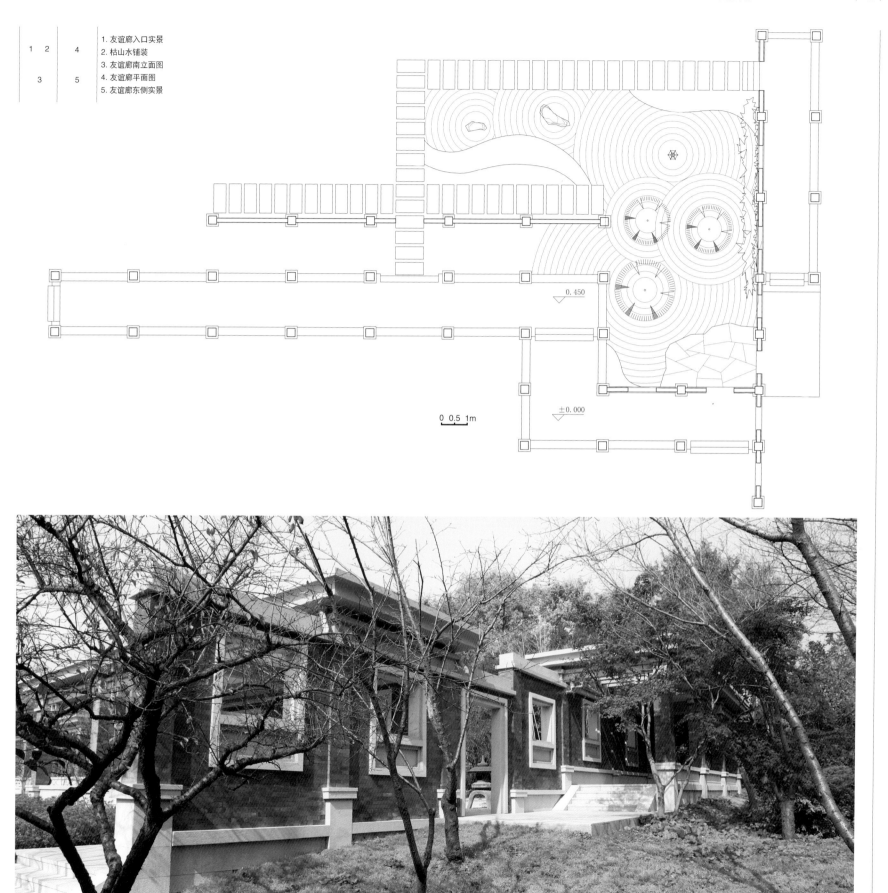

1. 友谊廊入口实景
2. 枯山水铺装
3. 友谊廊南立面图
4. 友谊廊平面图
5. 友谊廊东侧实景

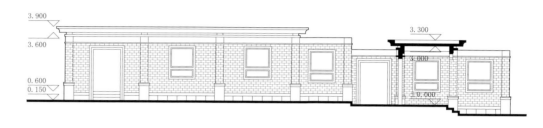
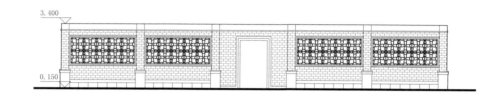
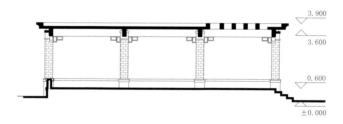
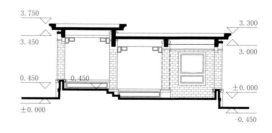

1. 友谊廊北侧实景
2. 友谊廊东立面图
3. 友谊廊剖面图一
4. 友谊廊北立面图
5. 友谊廊剖面图二
6. 友谊廊剖面图三

风景环境　SCENIC AREA LANDSCAPE

1. 友谊廊庭院秋景
2. 云蒸霞蔚框景
3. 园桥实景

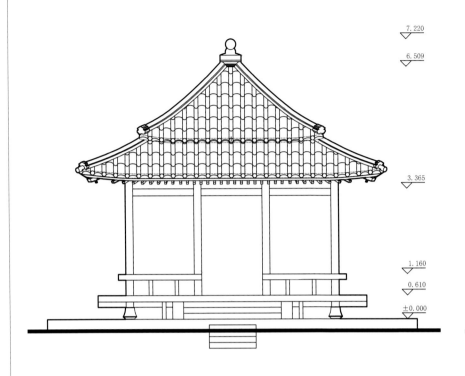

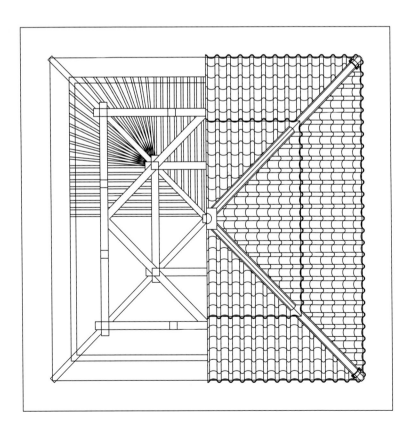
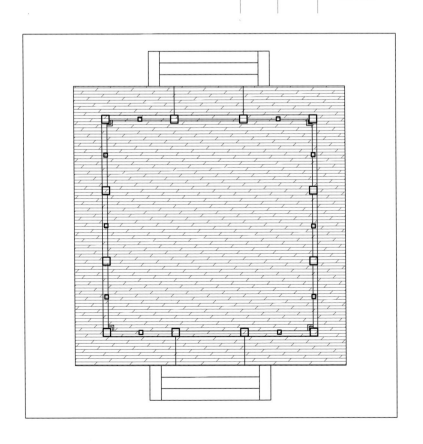

1. 观樱亭正立面图
2. 观樱亭侧立面图
3. 屋架仰视及顶平面图
4. 观樱亭平面图
5. 观樱亭实景

风景环境　SCENIC AREA LANDSCAPE

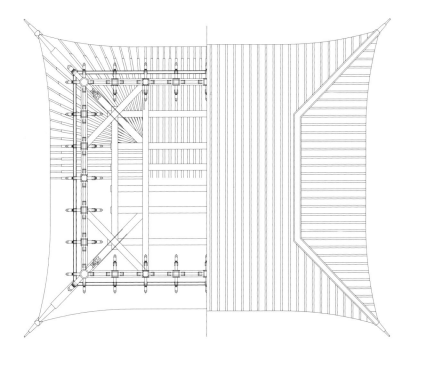

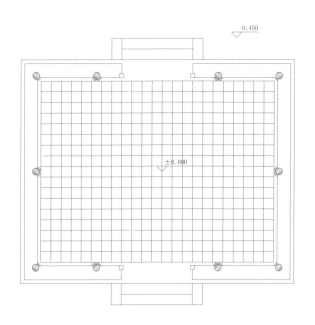

1. 胭脂雪卷棚歇山亭实景
2. 卷棚歇山亭侧立面图
3. 卷棚歇山亭剖面图一
4. 屋架仰视及顶平面图
5. 平面图
6. 卷棚歇山亭正立面图
7. 卷棚歇山亭剖面图二

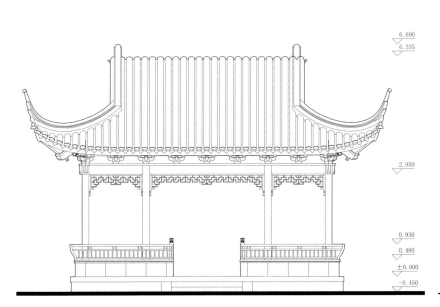

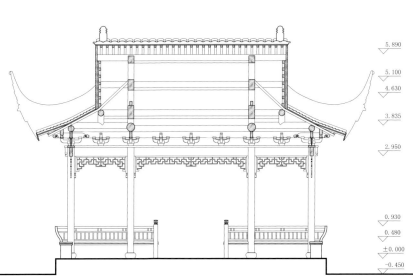

风景环境　SCENIC AREA LANDSCAPE

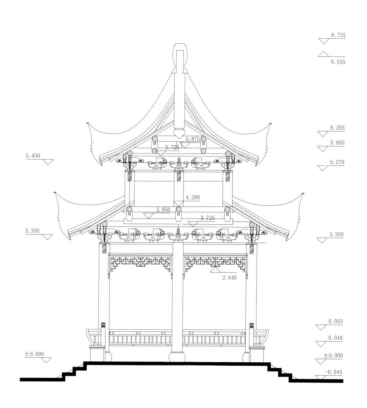

	2	1. 芳菲月亭实景
1		2. 芳菲月亭剖面图
	3	3. 芳菲月亭立面图

0　0.5　1m

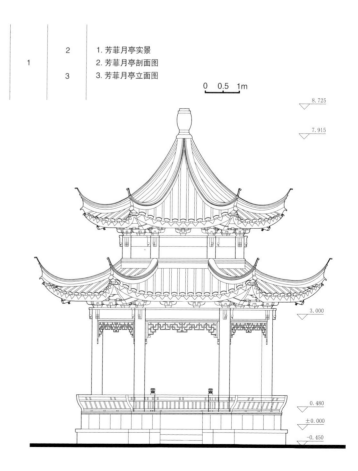

1. 樱花林鸟瞰实景
2. 芳菲月亭雪景
3. 芳菲月亭秋景

风景环境 SCENIC AREA LANDSCAPE

1. 紫藤廊实景
2. 紫藤廊立面图
3. 樱花林实景

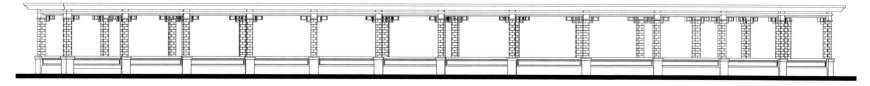

0 0.5 1m

南京市雨花台风景名胜区丁香花园设计
DESIGN OF DINGXIANG GARDEN IN YUHUATAI SCENIC SPOT, NANJING

项目地点：江苏省南京市	Location: Nanjing, Jiangsu	项目成员
设计时间：2011年	Design Period: 2011	景观设计：成玉宁、陈 思
建成时间：2012年	Completion Time: 2012	
委托单位：南京市雨花台管理局	Client: Nanjing Yuhuatai Authority	
用地面积：约1200平方米	Area: About 1200 m²	

项目概况

在南京雨花台烈士陵园，长眠着一位女烈士——白丁香，为了理想于二十二岁失去了自我及腹中胎儿的生命。丁香花园的景观设计围绕着白丁香烈士的故事展开，以"大爱无痕、大象无形"为创作理念。基地位于园区主轴线东侧一片西高东低的自然缓坡草地，面积约2000m²，东侧以无患子林为主，南侧为次生林地，中部为疏林草地，西侧为"忠魂颂"浮雕。在特定场所中生成的丁香花园是与场所精神对话的产物，它恬淡、质朴、宁静，融于周边环境，一片纯净的丁香林传达着独特的意境。

设计策略

1) 大爱无痕——景象的塑造

丁香花园由引导空间、瞻台、草坪、丁香林和弹洞涌泉5个部分构成。设计采用"留白"的艺术创作手法，有意识地借由一片平坦的草地从空间上将观者与纪念主体分离开来，以此形成纪念人物与观者之间的"距离"。西侧的"22株丁香树和一眼涌泉"是景观的主体，东侧是地势略低的木质"瞻台"。"瞻台"与"丁香林"之间产生的空间张力、略微的仰视，令观者的敬仰之情油然而生。

2) 大象无形——意境的生成

以"大象无形"为表达方式，22株丁香树象征着烈士短暂而绚丽的人生与精神的永恒。"弹洞涌泉"由烈士纪念馆装修剩余的93块汉白玉加工而成，静卧于草坡，象征着子弹洞穿大地而形成的"弹洞"，表达着对无情枪弹的憎恶。"弹洞"中汩汩涌出的清泉则似"血液"，寓意生命的不息。设计采用非线性叙事方式，突破时空结构的限定，通过拼接、打散与叠加组织景园空间，结合隐喻引发园林意境的生成。

丁香花园设计构思草图

Project Profile

In Nanjing Yuhuatai Martyrs Cemetery rests a female martyr named Bai Dingxiang, who, at 22 years old when she was pregnant, died for the sake of her great ideal. The landscape design of Dingxiang Garden focuses on martyr Bai Dingxiang's story, inspired by the idea of "the greatest love leaves no trace; the noblest image presents no form". The selected site is a slightly inclined natural meadow located in the east of the main axis. With its west land higher than the east, the site covers an area of 2000m². Soapberry trees mainly grow in the east, secondary forest in the north, open forest and grassland in the central, and a relief sculpture named "Ode to Loyal Soul "stands in the west. Dingxiang Garden is a spiritual product created on the specific site. Tranquil, plain and quiet, this garden assimilates itself into surroundings, and expresses a unique artistic conception through its pureness.

Design Strategy

1)The greatest love leaves no trace—the creation of artistic conception.

Dingxiang Garden consists of five parts, namely a guiding space, an observation deck, lawn, Diangxiang woods and the Bullet-hole Spring. The design uses the artistic creation method—"blank-leaving", that is, utilizing a flat meadow to separate the viewer and the memorial subject from spatial perspective, thus creating a "distance" between the memorial subject and the viewer. Located in the west is the main landscape, "22 Dingxiang trees and the Singe-hole Spring", while in the east where the land is relative lower stands the wood observation deck. It's expected that spectators'admiration will arise spontaneously from the spatial tension produced between "Observation Deck "and "Dingxiang Woods" and the slightly upward angle they take when they look up from the deck.

2)The noblest image presents no form—the creation of artistic conception.

Centered on the idea that "the noblest image presents no form", 22 clove trees symbolize the martyr's short but gorgeous life and her everlasting spirit. Lying on the meadow, "Bullet-hole Spring" is made of 93 pieces of white marble left by the fitting-out works of the martyr memorial museum. The "bullet-hole" stands for the image that a bullet shoots through the land, expressing hatred to those indifferent guns and bullets. Clear spring continuously gushing from the "bullet-hole" is like "blood", which implies that life is endless. The design adopts nonlinear narration to break through the limits of the temporal-spatial structure. Through blending, scattering and overlapping to reorganize landscape space, the design combines metaphor to create the artistic conception.

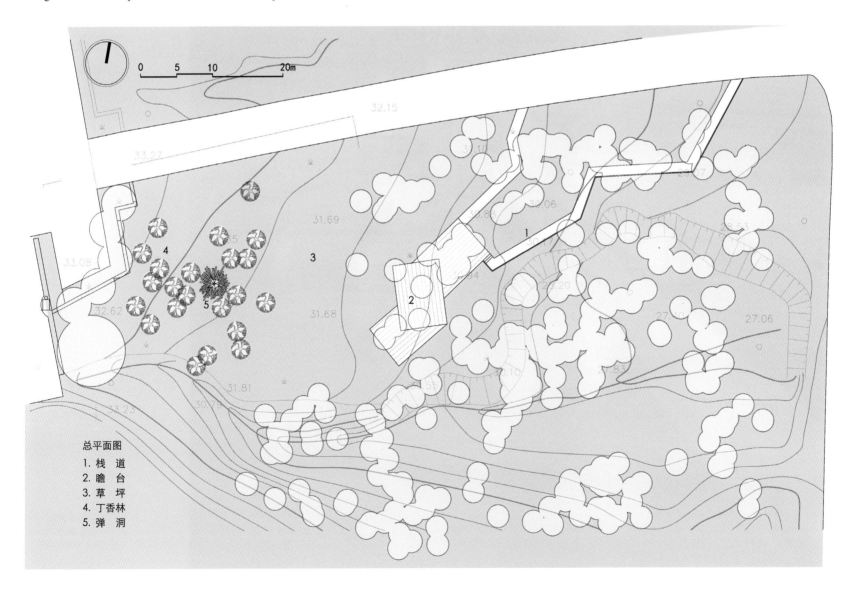

总平面图
1. 栈 道
2. 瞻 台
3. 草 坪
4. 丁香林
5. 弹 洞

1. 弹洞涌泉
2. 瞻台实景
3. 缅英林实景

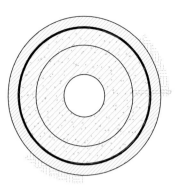
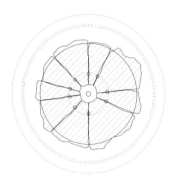

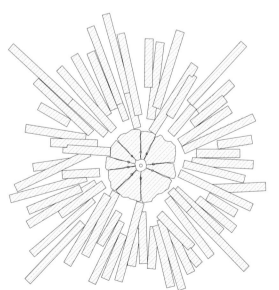
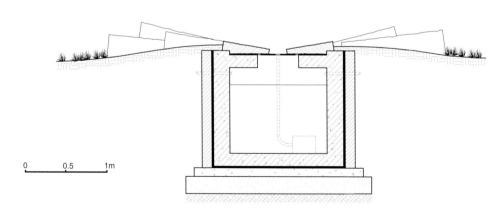

1	2
3	4
5	

1. 弹洞涌泉近景
2. 细部详图
3. 平面图
4. 剖面图
5. 弹洞涌泉实景

| 1 | 2 | 3 | 1. 林间栈道
2. 林下瞭台
3. 缅英林实景

风景环境 SCENIC AREA LANDSCAPE | 67

花园瞻台及栈道实景

南京市牛首山风景区（北部景区）景观及建筑设计
LANDSCAPE AND ARCHITECTURE DESIGN OF NORTHERN NIUSHOUSHAN SCENIC SPOT, NANJING

项目地点：江苏省南京市	Location: Nanjing , Jiangsu	项目成员
设计时间：2013 年	Design Period: 2013	景观设计：成玉宁、袁旸洋、汪瑞军
建成时间：在建	Completion Time: Under Construction	建筑设计：成玉宁、方 颖、吴雪鎏、穆燕洁
委托单位：南京软件谷文化旅游发展有限公司	Client: Nanjing Software Valley Cultural Tourism Development Co., Ltd.	结构设计：盛春陵、周毅雷
用地面积：60.50 公顷	Area: 60.50 ha	水电设计：王晓晨、李滨海
		雕塑方案：成玉宁
		雕塑深化：钱大经

项目概况

牛首山风景区是南京历史上南北轴线的重要节点，其中北部景区面积约为 60.72 公顷。设计充分挖掘场所的历史文化内涵，将多元的佛文化体验融入一条蜿蜒于自然山水之间的"寻禅道"之中，营造出"天人合一"的禅宗境界。

设计策略

设计以"天阙礼佛顶，山水寻禅机"为概念构思，利用自然的地形地貌，构建"起、承、转、合"的空间结构。

1）文化表达策略

通过佛教造像、佛文化符号彰显、禅宗故事再现、佛事活动体验及寻禅修禅等多种方式，对佛文化主题进行景观化的表达。设计将禅宗的思想精神融入山林、水系、雕塑、建筑与小品之中，结合禅宗养生、饮食、艺术等特色游览项目，创造出丰富的文化体验形式。

2）生态优化策略

设计依据地形地貌及汇水条件对原有水系进行梳理，提升暴雨时期场地的调蓄能力，为日常景观灌溉提供充足的水源。同时，优化生态环境，结合地带性适生植物配置，建构典型生境特征。

3）空间营造策略

从场所空间形态中提取出"山、谷、溪"三大要素，设计紧密结合原有场地竖向及空间特色，依两翼山势，傍蜿蜒溪水，沿谷底依次错落布置景观节点。景区入口处的空间壮阔开敞，序列性、标志性突出；进入"寻禅道"空间迂曲委婉、幽深宁谧，形成步移景异的空间变化。

Project Profile

Niushoushan Scenic Spot is an important node at which the northern and southern axes intersect in Nanjing's history, and the north part covers an area of 60.72 hectares. The design fully extracts the site's historical and cultural connotation, and integrates various Buddhist culture experience into a "Buddhist Searching Road" winding along the natural landscape, thus creating the Buddhist realm of "harmony between man and nature."

Design Strategy

Inspired by the concept that "since the heaven pay tribute to Foding Palace, only among mountains and rivers can one find the genuine Buddhist", the design utilizes the natural landform to build a spatial structure with "opening, developing, changing and concluding" units.

1) Culture expressing strategy

The Buddhist culture theme is expressed by landscapes in various ways, such as Buddhist statues, highlighting Buddhist culture symbols, representing Buddhist stories, experiencing Buddhism activities, searching and practicing Buddhist. The design integrates the spirit of Buddhist ideas into mountains, forests, water systems, sculptures, architectures and accessorial buildings, and combines distinctive tour events like art, food and health maintenance in line with Buddhist ideas, thus creating a variety of cultural experiences.

2) Ecology optimizing strategy

In line with the landform and catchment condition, the design sorts out the original water system, enhances the site's regulation capacity for rainstorm periods, and manages to provide abundant water to daily landscape irrigation. Meanwhile, the design optimizes the eco-environment and arranges plants in suitable living sections, successfully establishing typical habitat characteristics.

3) Space creating strategy

Three major elements are extracted among the site's spatial form, namely "mountain, valley and stream". Integrating the original site's vertical and spatial features, the design sequentially arranges landscape nodes near the mountain, by the river and alongside the valley floor. The space near the entrance is open and broad, showing an outstanding sequence and order. After entering the "Buddhist Searching Road", the space becomes zigzag and tranquil, and scenes change with viewer's steps.

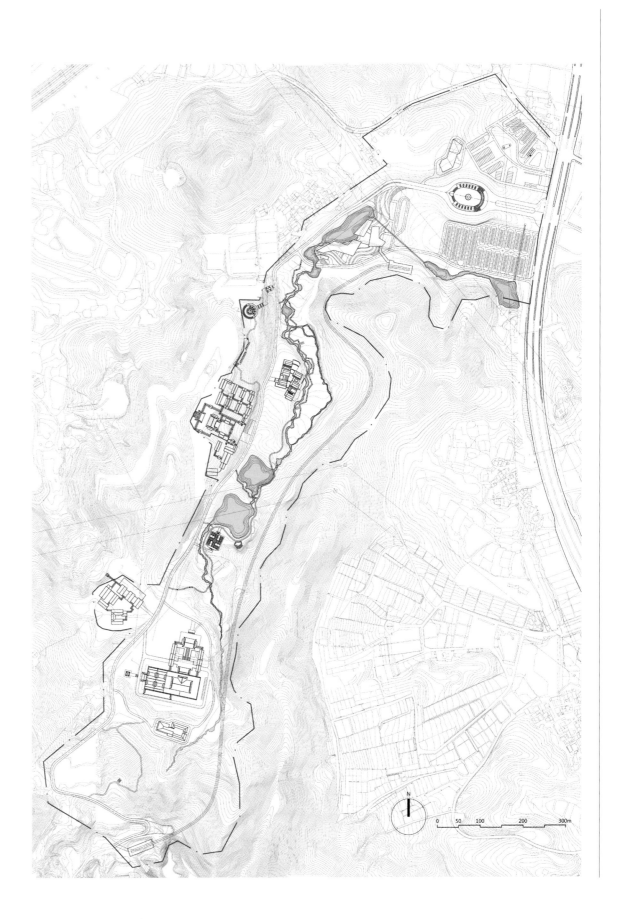

牛首山北部景区寻禅道平面图

入口景观实景

1. 万象更新设计草图
2. 万象更新视线分析图
3. 万象更新剖面图
4. 万象更新平面图

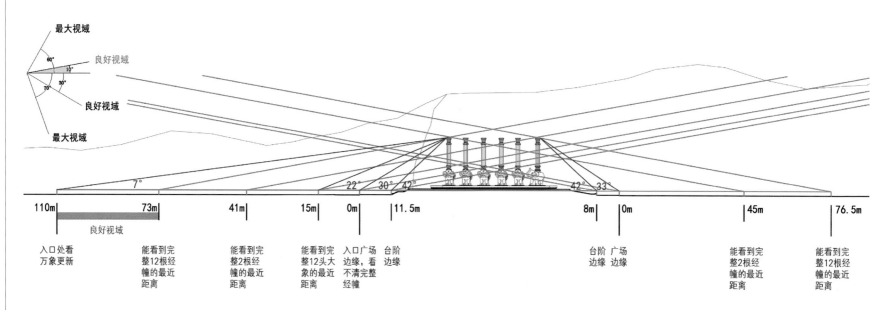

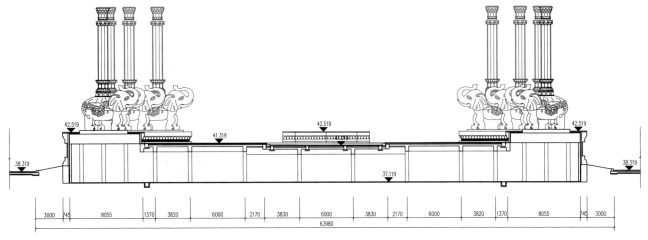

万象更新B-B剖面 1:300

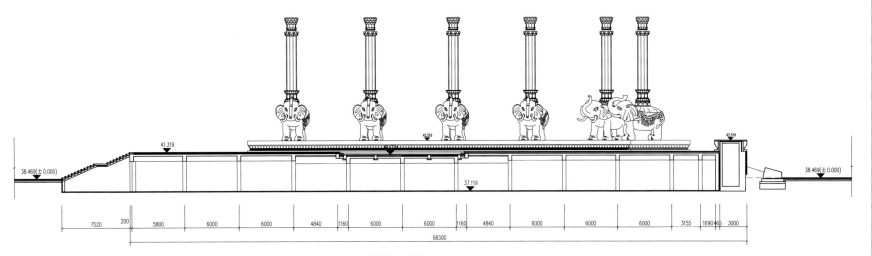

万象更新A-A剖面 1:300

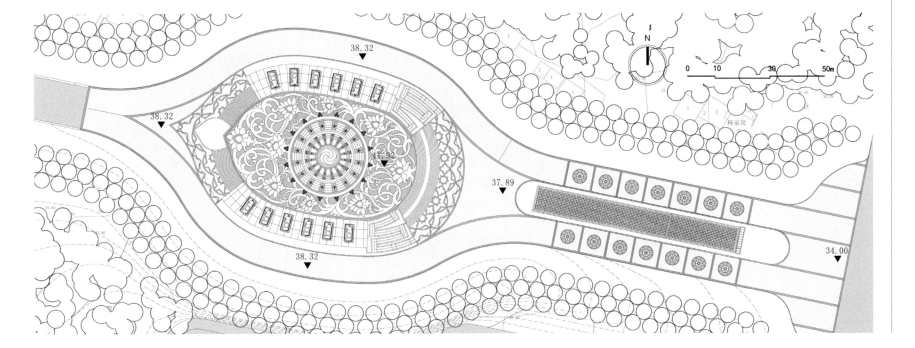

万象更新夜景效果图

1. 万象更新广场实景
2. 万象更新东侧雕塑实景
3. 万象更新西侧雕塑实景
4. 镜面水池倒影实景

风景环境 SCENIC AREA LANDSCAPE

万象更新西侧实景

涤心潭实景

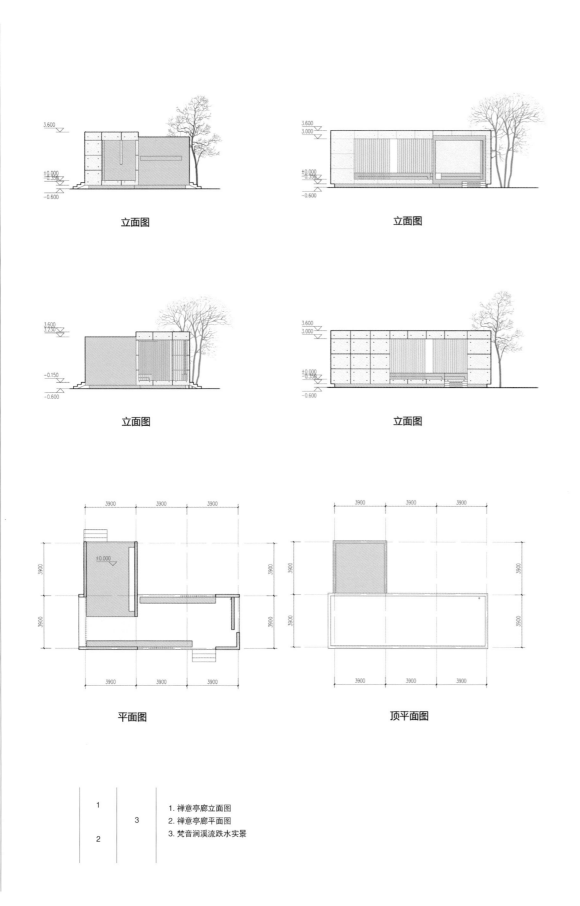

1	3	1. 禅意亭廊立面图
2		2. 禅意亭廊平面图
		3. 梵音涧溪流跌水实景

风景环境 SCENIC AREA LANDSCAPE | 85

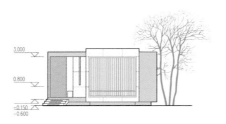

立面图

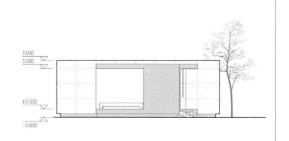

立面图

立面图

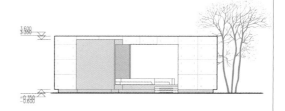

立面图

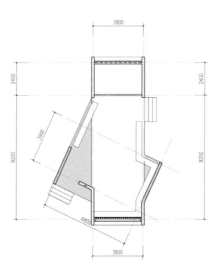

平面图

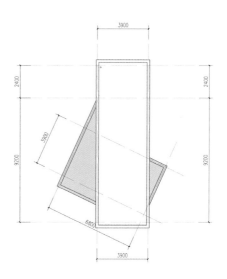

顶平面图

	2	1. 塔影湖实景
1		2. 禅意亭廊立面图
	3	3. 禅意亭廊平面图

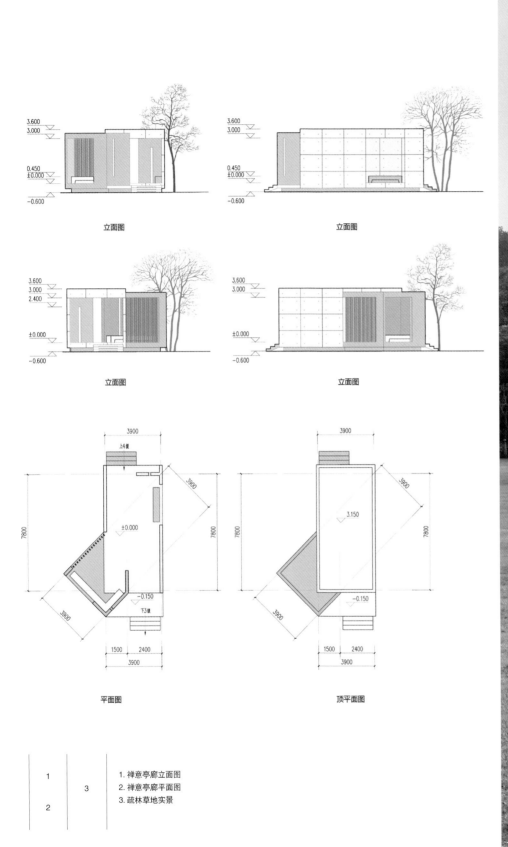

1. 禅意亭廊立面图
2. 禅意亭廊平面图
3. 疏林草地实景

| 88 | SCENIC AREA LANDSCAPE 风景环境

1. 疏林草地实景
2. 涤心潭夜景
3. 梵音涧溪流实景

1. 拈花微笑方案
2. 观光缆车站鸟瞰图
3. 观光缆车站方案

风景环境 SCENIC AREA LANDSCAPE

| 92 | SCENIC AREA LANDSCAPE   风景环境

1	2	7
3	4	8
		9
5	6	10

1. 游客及管理中心一层平面图　　6. 游客及管理中心 B-B 剖面图
2. 游客及管理中心二层平面图　　7. 游客及管理中心南立面图
3. 游客及管理中心三层平面图　　8. 游客及管理中心北立面图
4. 游客及管理中心地库平面图　　9. 游客及管理中心西立面图
5. 游客及管理中心 A-A 剖面图　　10. 游客及管理中心东立面图

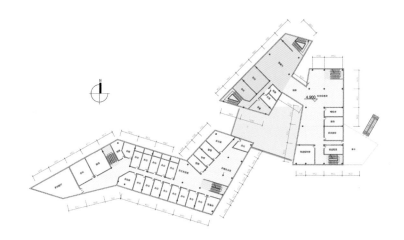
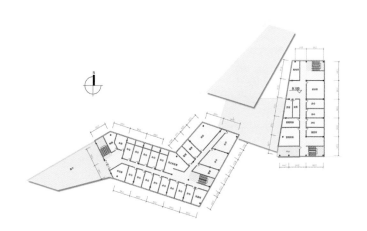
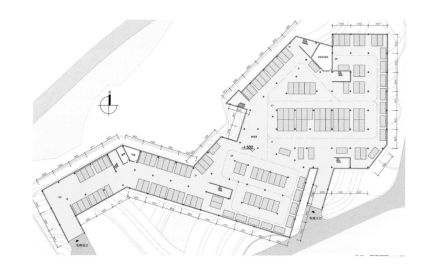

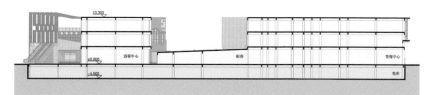

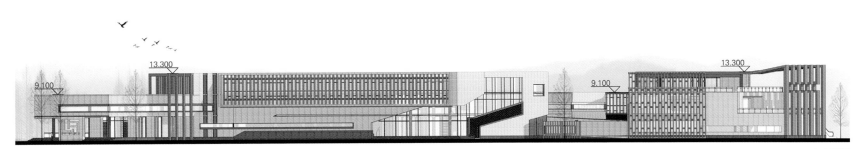

南立面图

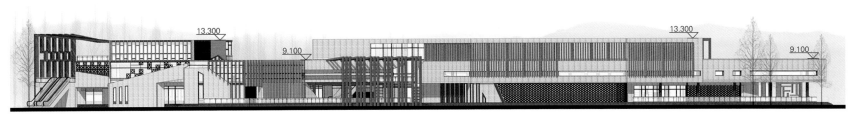

北立面图

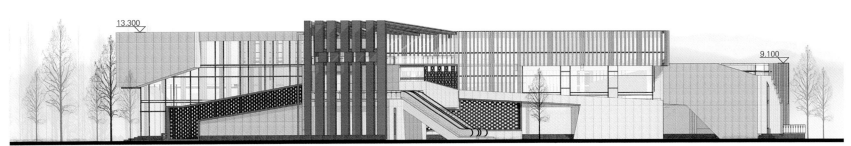

西立面图

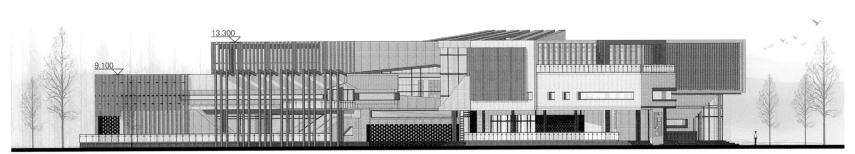

东立面图

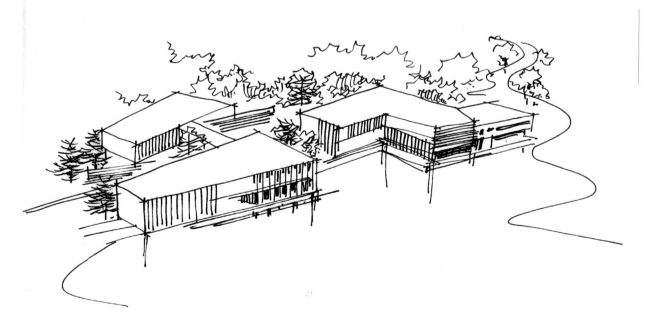

1	3
2	4

1. 游客及管理中心鸟瞰图
2. 游客及管理中心设计草图
3. 游客及管理中心效果图
4. 游客及管理中心夜景效果图

风景环境 SCENIC AREA LANDSCAPE | 95

常熟市沙家浜东扩湿地景观与建筑设计
SHAJIABANG WETLAND ARCHITECTURE AND LANDSCAPE DESIGN, CHANGSHU

项目地点：江苏省常熟市	Location: Changshu, Jiangsu	项目成员
设计时间：2008年	Design Period: 2008	景观设计：成玉宁、许立南、徐悦
建成时间：2010年	Completion Time: 2010	建筑设计：成玉宁、张亚伟、许立南、成实、戴丹骅、黄妙玲
委托单位：沙家浜风景区管委会	Client: Shajiabang Scenic Area Management Committee	结构设计：王伟成
用地面积：162700平方米	Area: 162700 m²	水电设计：刘俊、王晓晨

项目概况

常熟沙家浜湿地公园东扩工程总建设面积为16.27万平方米，基地背临阳澄湖，纵横交错的河港和茂密的芦苇营造出形态多变的水陆景观空间。设计以京剧样板戏《沙家浜》为文化背景，着力提升了场所的游览趣味和景观品质，塑造了生态自然的沙家浜特有湿地景观。

设计策略

设计因地制宜地梳理水系，恢复湿地生境，巧妙安排景观展示、文化展示以及其他参与性项目，并依据记载恢复历史建筑"云庆庵"，营造出和谐、生态的景观环境。

1）湿地景观的生活化

湿地植物品种园、湿地生态示范园和芦荡人家是设计的三个重要组成部分。西北的湿地植物品种园是展示江南地区湿地植物的专类园。中部以游憩强度较低的农业采摘、渔业体验与湿地探索功能为主。东南为湿地生态示范园，东部片区设立湿地生态恢复区，在展示湿地生态链和湿地动植物群落的同时，设有水街、芦荡人家等展现湿地人居风情。

2）湿地生境营造

设计构建了典型江南水乡湿地生态环境，通过梳理水系、拓宽湿地滩面、增加水体岸线长度并调整岸线形态，形成了从陆地到湖面的自然缓坡湿地景观。同时根据陆地、水体不同标高的界面和季节性水位变化，配植以沉水、浮水、挺水植物。

3）季相造景策略

设计旨在营造"芦花放、稻谷香、岸柳成行"的江南水乡意境。在不同的地块中分别描绘"春是桃红柳绿，夏是荷藕飘香，秋是杏林尽染，冬是雪融芦花"的四季景象。

Project Profile

The total construction area of the eastward expansion project of Changshu Shajiabang Wetland Park is 162,700m². The construction site is at the back of Yangcheng Lake, where crisscross rivers and harbors as well as dense reeds create a variety of land and water landscape spaces. With Shajiabang, the model opera of Peking opera as the cultural background, the design makes efforts in enhancing the site's touring and landscape quality, and portraying an ecological and natural wetland landscape that is peculiar to Shajiangbang.

Design Strategy

The design sorts out water system according to the local conditions, restores the wetland habitat, and ingeniously arranges the display of landscape and culture as well as other participatory events. It also restores the historic architecture Yunqing Nunnery based on achieves, thus creating a harmony and ecological landscape environment.

1) Make wetland landscape closer to life

The design consists of three main parts, namely Wetland Botanic Species Garden, Wetland Ecological Demonstration Garden and Ludang Renjia Resort. Located in the northwest, Wetland Botanic Garden is a specialized garden to exhibit wetland plants that usually grow in regions south of the Yangtze River. The central section is a site for activities of lower intensity, such as crop harvesting, fishery experiencing and wetland exploration. In the southeast stands the Wetland Ecological Demonstration Garden, and to the east of the central section sets the Wetland Ecological Restoration Zone. While the wetland's ecological chain as well as flora and fauna community is displayed, the design also shows its local living customs through setting Water Streets and the Ludang Renjia Resort.

2) The construction of wetland habitat

The design establishes typical wetland eco-environment of Yangtze River delta. By sorting out water system, expanding wetland beach face, lengthening water front and adjust shoreline form, the design creates a wetland landscape with a natural gentle slope extending from land to the lake. Meanwhile, based on the elevation difference between land and water as well as the seasonal changing water level, submerged plants, floating plants and emergent plants are planted accordingly.

3) The strategy of landscaping with seasoning characteristics

The design is aimed at creating an artistic conception peculiar to regions of Yangtze delta, where reed catkins bloom, paddy gives off fragrance and willows grow alongside the river banks. In different sections, landscape of four seasons is portrayed, like the ancient Chinese sentence: "In spring, red peach blossoms bloom and green willows billow; in summer, fragrance of lotus roots goes with the wind; in fall, apricot woods take on golden coats; in winter, reed blossoms sleep under the snow."

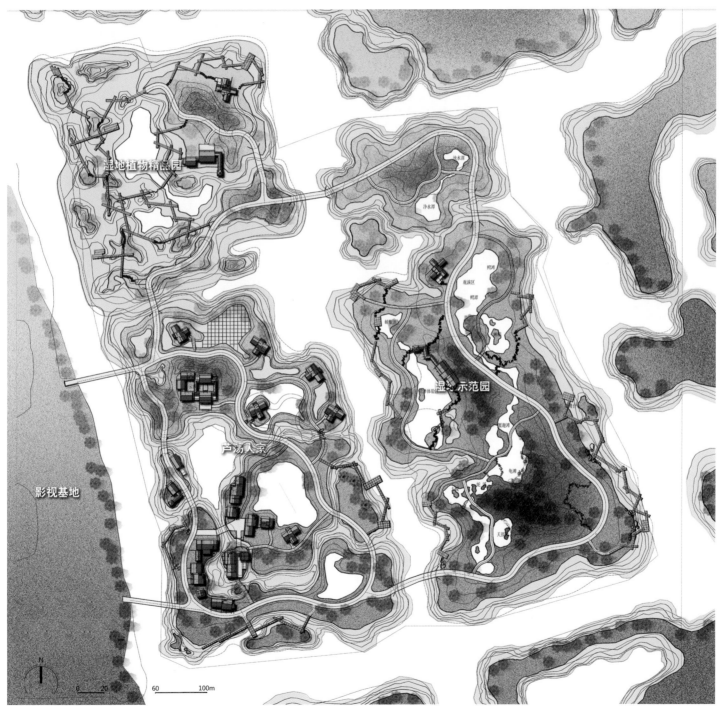

沙家浜东部湿地总平面图

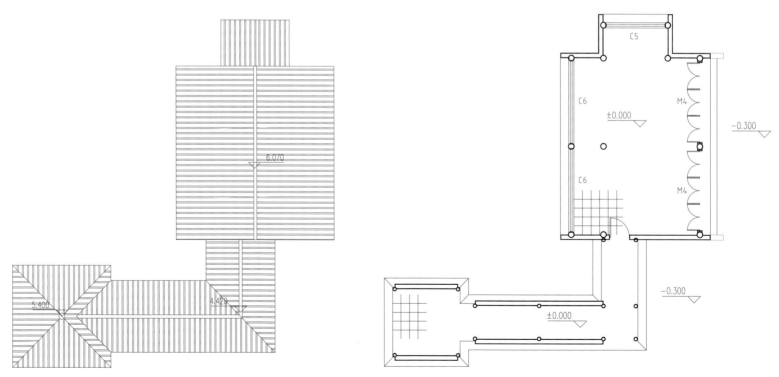

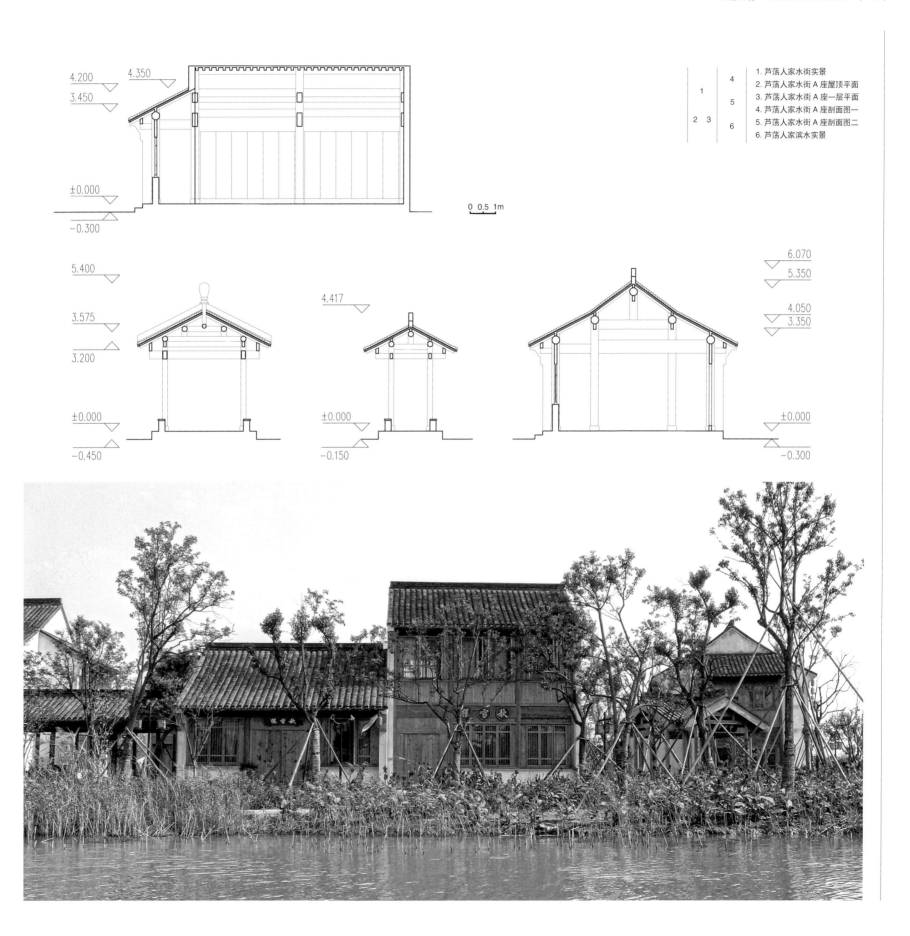

1. 芦荡人家水街实景
2. 芦荡人家水街A座屋顶平面
3. 芦荡人家水街A座一层平面
4. 芦荡人家水街A座剖面图一
5. 芦荡人家水街A座剖面图二
6. 芦荡人家滨水实景

1. 芦荡人家内街实景
2. 芦荡人家雪景
3. 短凳桥雪景

风景环境　SCENIC AREA LANDSCAPE

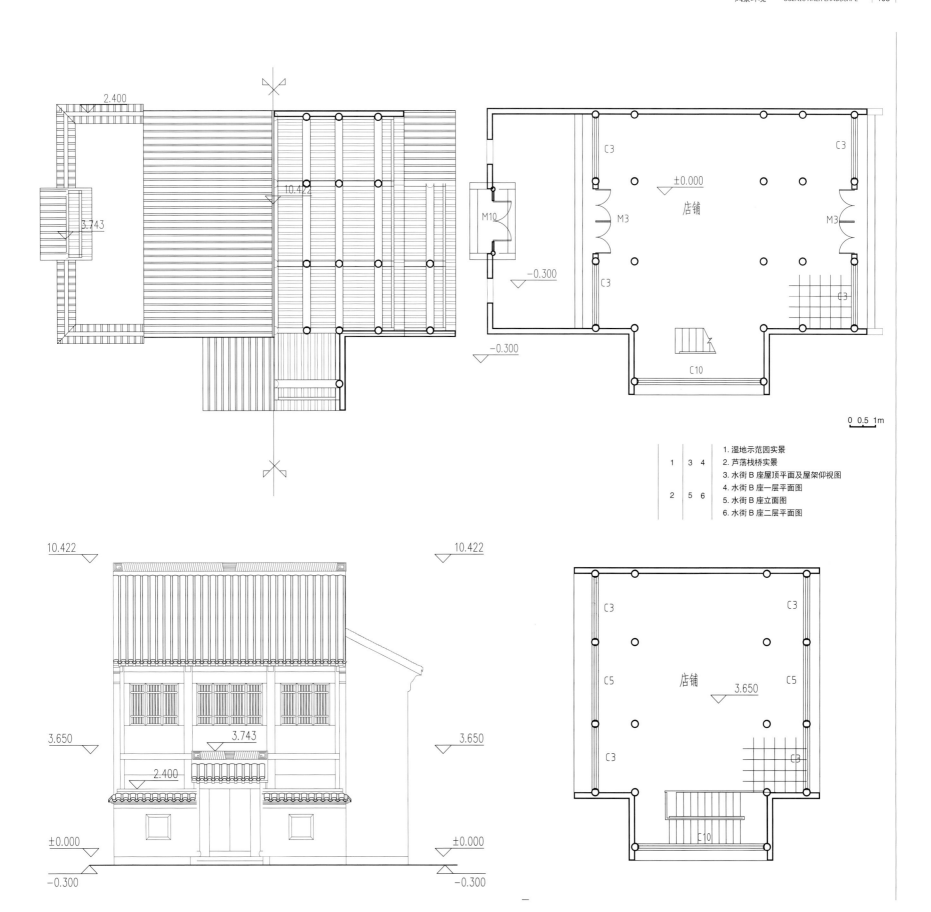

1. 湿地示范园实景
2. 芦荡栈桥实景
3. 水街B座屋顶平面及屋架仰视图
4. 水街B座一层平面图
5. 水街B座立面图
6. 水街B座二层平面图

1. 芦荡人家廊亭实景
2. 芦荡人家水街实景
3. 云庆庵平面图

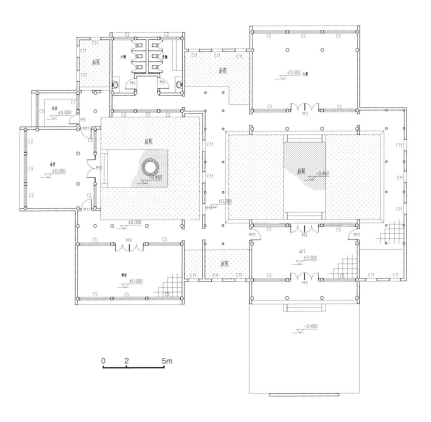

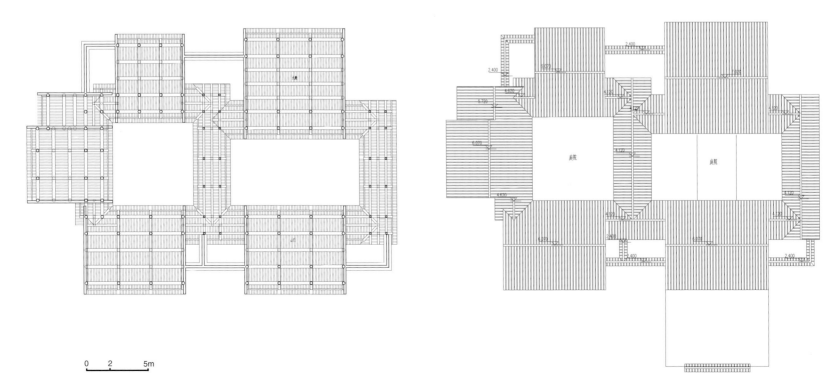

0　2　5m

1	4	
	5	
2 3	6	

1. 芦荡人家木栈桥实景
2. 云庆庵屋架仰视图
3. 云庆庵屋顶平面图
4. 云庆庵立面图
5. 云庆庵剖面图
6. 湿地栈桥雪景

0 2 5m

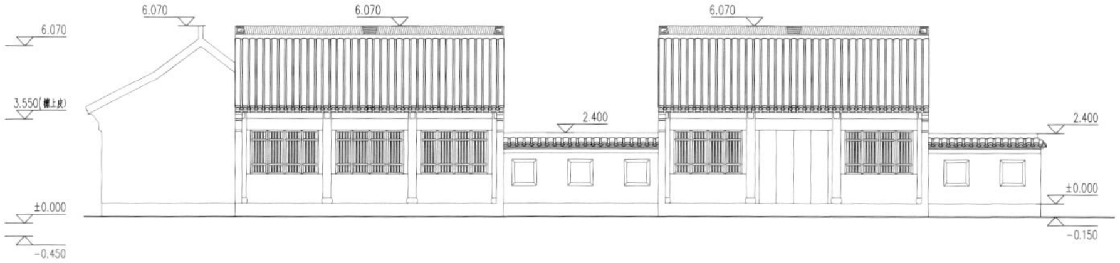

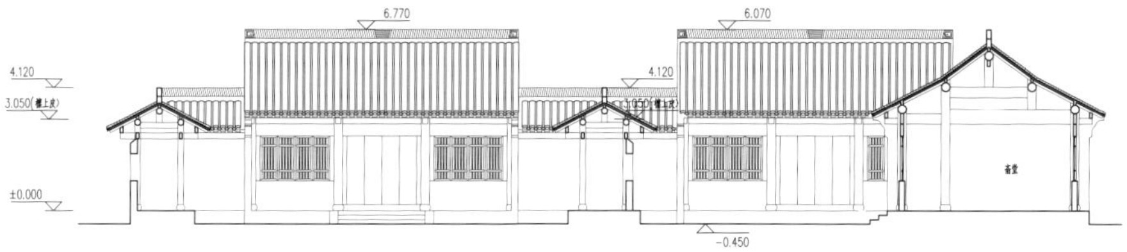

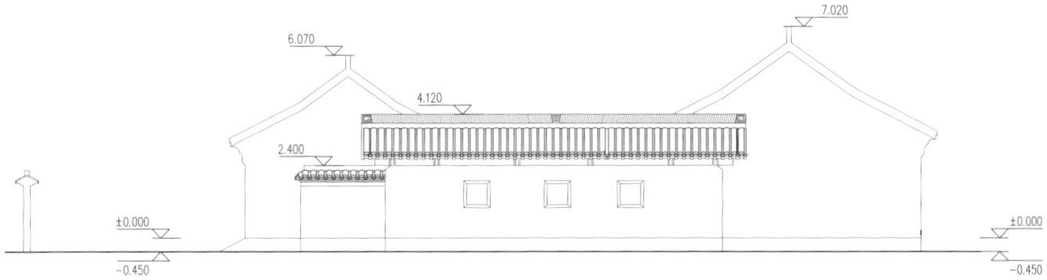

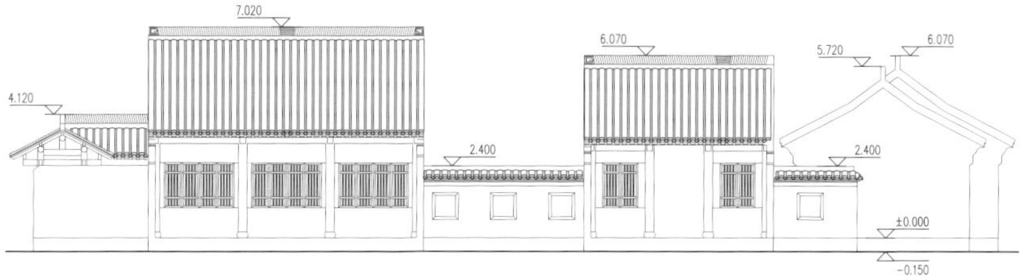

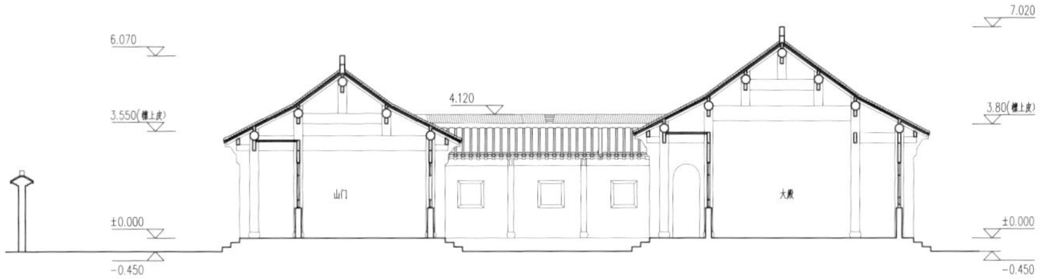

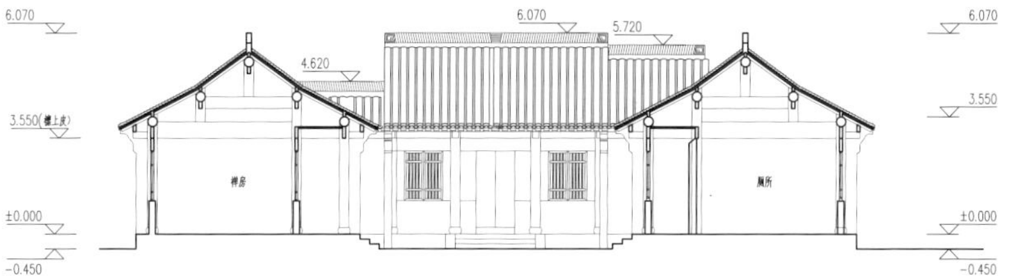

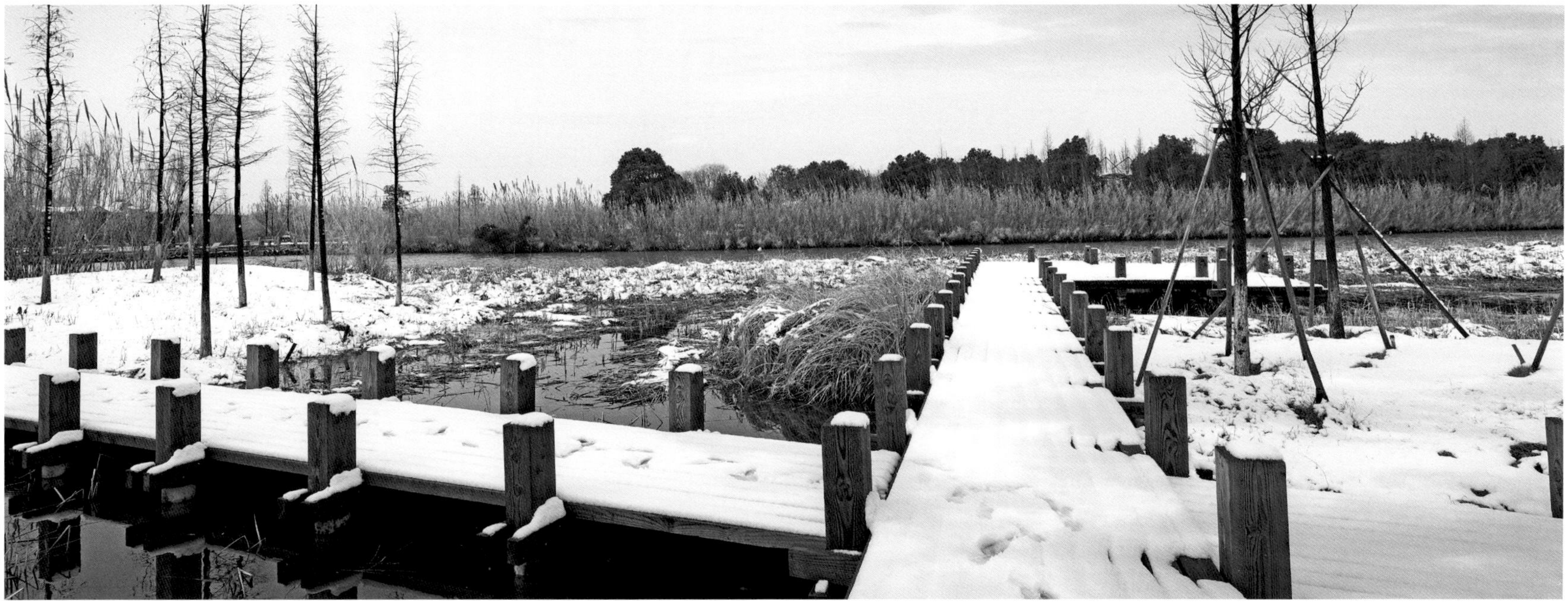

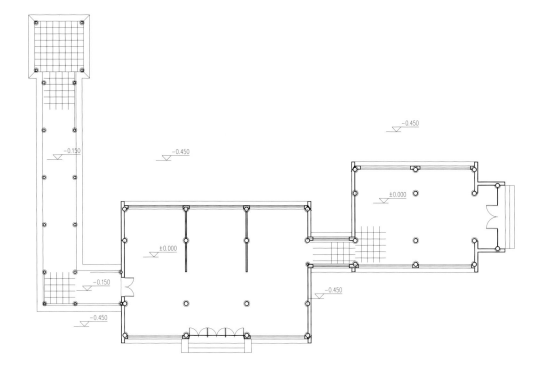
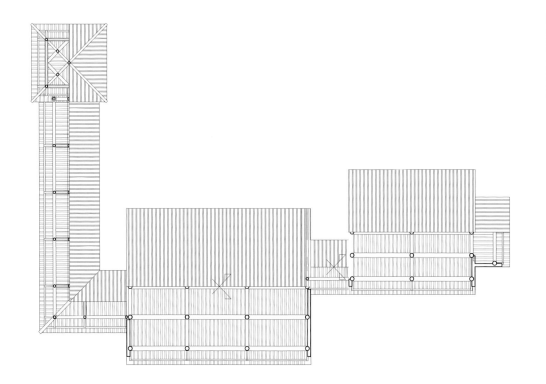

1. 湿地植物精品园展馆平面图
2. 屋顶平面及屋架仰视图
3. 建筑东立面图
4. 建筑剖面图
5. 建筑北立面图
6. 展馆栈桥实景

风景环境 SCENIC AREA LANDSCAPE | 109

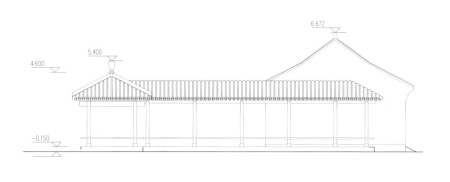
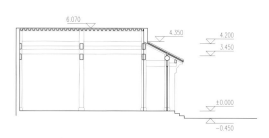
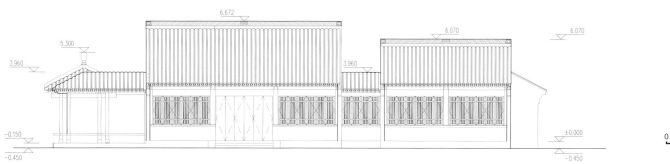
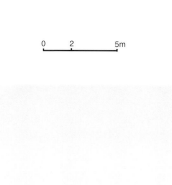

| 110 | SCENIC AREA LANDSCAPE　　风景环境

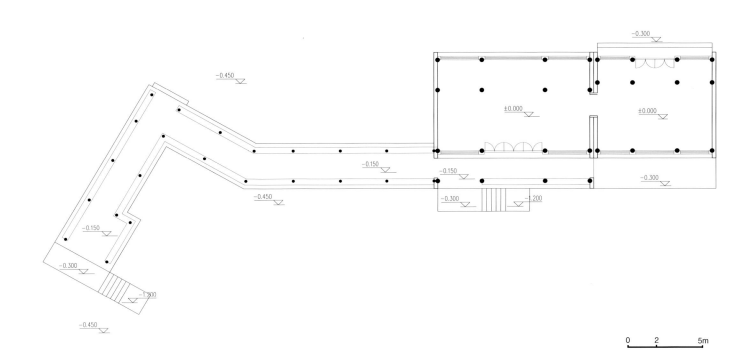

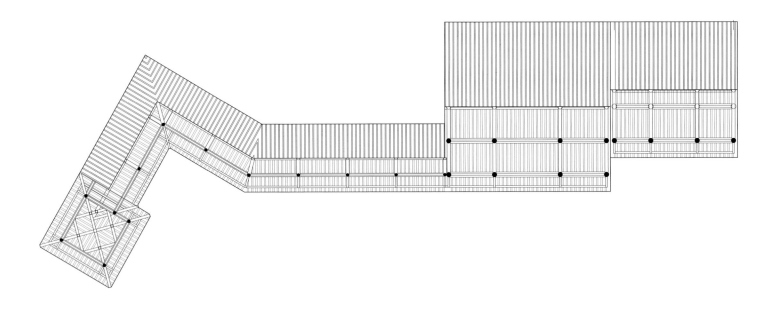

| 1 | 3 |
| 2 | 4 |

1. 渔乐轩平面图
2. 屋顶面及屋架仰视图
3. 渔乐轩立面图
4. 渔乐轩雪景

风景环境　SCENIC AREA LANDSCAPE | 111

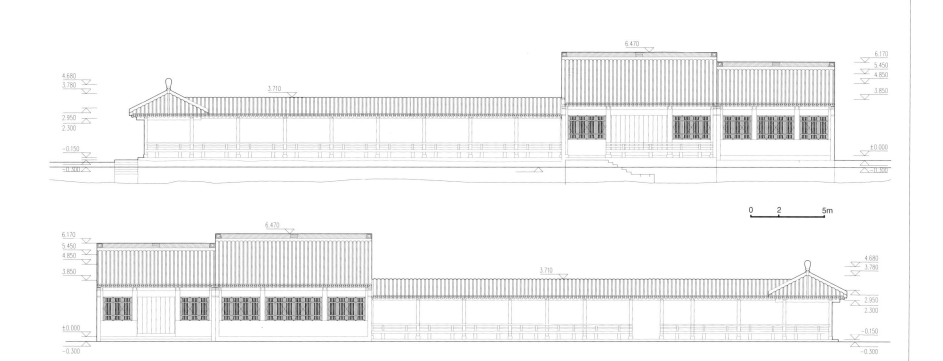

风景环境　SCENIC AREA LANDSCAPE　| 113 |

芦苇荡实景

| 1 | 2 | 1. 湿地雪景
2. 迁建古桥实景

城市公园
URBAN PARK

南京市石头城遗址公园概念性规划设计
CONCEPTUAL PLANNING AND DESIGN OF SHITOUCHENG HERITAGE PARK, NANJING

项目地点：江苏省南京市	Location: Nanjing, Jiangsu	项目成员
设计时间：2013年	Design Period: 2013	景观设计：成玉宁、周聪惠、谭 明、赵玉龙、濮岳川、刘 悦、谢明坤
建成时间：在建	Completion Time: Under Construction	建筑设计：成玉宁、谭 明、单梦婷、孟 梦
委托单位：南京市旅游委员会	Client: Nanjing Tourism Committee	
用地面积：107.5公顷	Area: 107.5 ha	

项目概况

东汉建安十七年（公元212年）孙权在金陵邑旧址上修筑"石头城"，是南京作为"六朝古都"的发端。经考古证实，石头城遗址的范围涵盖现南京清凉山公园、国防园以及乌龙潭公园。设计通过整合、提升、改造现有三个城市公园，打造出一个集遗址保护与展示、文化旅游、教育科研、生态休闲等多功能为一体的大遗址文化公园。

设计策略

基于场所的生态敏感性和建设适宜性分析，划定保护区域；结合考古发现，尊重历史文化，依对象的不同采取保护、复建、演绎等设计策略。根据需要复建清凉台、清凉寺、兰苑等景观节点，从而完善、提升园区游览功能。

设计通过研究视线通廊及竖向关系，分析通视的可能性，发掘景观节点之间的空间对位关系。以六朝文化为主，设计中兼顾佛教文化、国防文化、书画文化、名人文化等特色文化，实现文化的多元融合。

按照规模、年代、性质类型，采用景观的手法对考古遗址进行保护性展示，直观地呈现历史状态。通过绿道和游线的设计加强与秦淮河、明城墙等文化景观资源的联系，放大了遗址公园的空间效应。

Project Profile

In the 17th year since Emperor Jian'an ruled the Eastern Han Dynasty (A.D. 212), Sun Quan built Shitoucheng (literally Stone City) on the former site of Jinglingyi. This is the beginning when Nanjing became the capital of six ancient dynasties. Confirmed by archaeological studies, the relics of Shitoucheng ranged from Nanjing Qingliangshan Park, Nanjing National Park to Nanjing Wulongtan Park. Through integrating, improving and reconstructing the existing three city parks, this design constructs a great multifunctional heritage park which meets the demands for protecting and displaying relics, cultural tourism and ecological recreation.

Design Strategy

Based on analyzing the site's ecological sensitivity and constructional suitability, the protection area is designated; in line with archeological findings, this design pays its respect to history and culture through taking corresponding design strategies for different objects, such as protection, reconstruction, and deduction. According to the need, some landscape nudes have been reconstructed, including Qingliang Terrace, Qingliang Temple, and Lanyuan Garden, which further improves the touring function of the park.

The design analyzes the possibility of inter-visibility and makes out the spatial para-position between landscape nudes through studying the sight corridor and vertical space. Focusing on cultures of six dynasties, the design integrates diverse cultures, including Buddhist culture, national defense culture, calligraphy and painting culture, as well as celebrity culture.

According to the site's scale, built time, properties and types, the design exhibits archaeological sites in a protective manner to visually demonstrate their historical conditions. The design of greenway and tour routes strengthens the association between cultural landscape resources, such as the Qinhuai River and the Ming City Wall, and amplifies the spatial effect of the heritage park.

清凉台设计构思草图

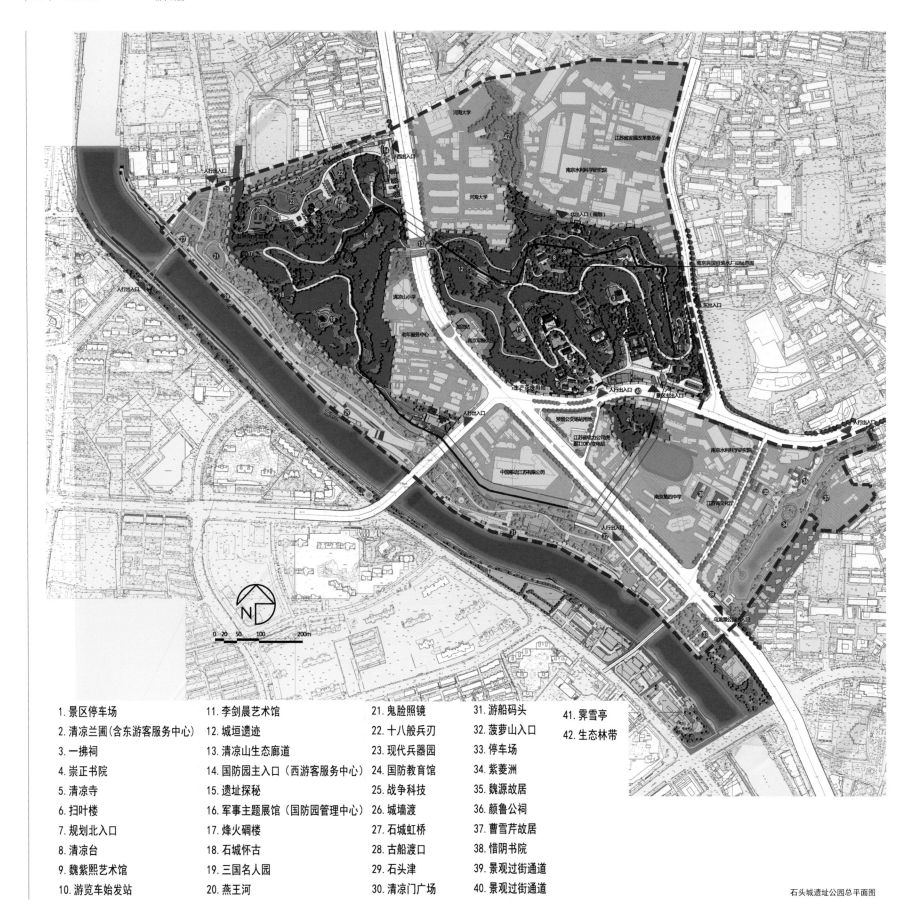

1. 景区停车场	11. 李剑晨艺术馆	21. 鬼脸照镜	31. 游船码头	41. 霁雪亭
2. 清凉兰圃(含东游客服务中心)	12. 城垣遗迹	22. 十八般兵刃	32. 菠萝山入口	42. 生态林带
3. 一拂祠	13. 清凉山生态廊道	23. 现代兵器园	33. 停车场	
4. 崇正书院	14. 国防园主入口(西游客服务中心)	24. 国防教育馆	34. 紫菱洲	
5. 清凉寺	15. 遗址探秘	25. 战争科技	35. 魏源故居	
6. 扫叶楼	16. 军事主题展馆(国防园管理中心)	26. 城墙渡	36. 颜鲁公祠	
7. 规划北入口	17. 烽火碉楼	27. 石城虹桥	37. 曹雪芹故居	
8. 清凉台	18. 石城怀古	28. 古船渡口	38. 惜阴书院	
9. 魏紫熙艺术馆	19. 三国名人园	29. 石头津	39. 景观过街通道	
10. 游览车始发站	20. 燕王河	30. 清凉门广场	40. 景观过街通道	

石头城遗址公园总平面图

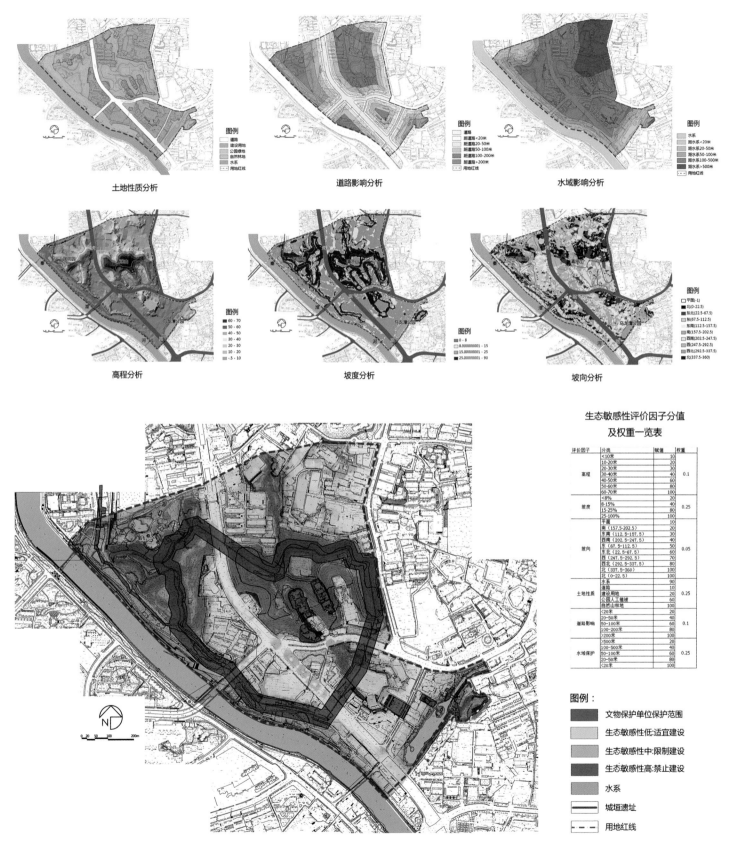

石头城遗址公园 ARCGIS 生态适宜性分析

| 120 | URBAN PARK | 城市公园

图 例:
- 公共开放空间
- 城垣遗址保护与展示环
- 秦淮河-明城墙风光带
- 游憩活动中心
- 景观视线高点
- 城市空间协调区
- 城市空间协调界面
- 项目红线
- 景观对景视线

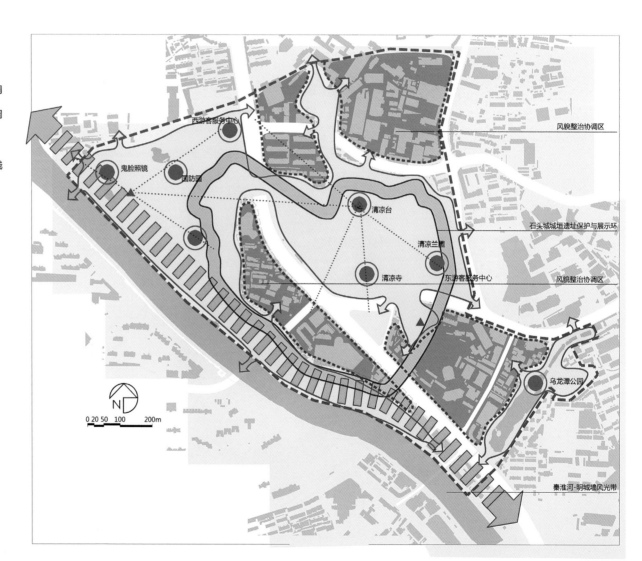

1. 城市环境、密度、建筑高度分析
2. 空间结构规划图
3. ARCGIS 空间视域分析

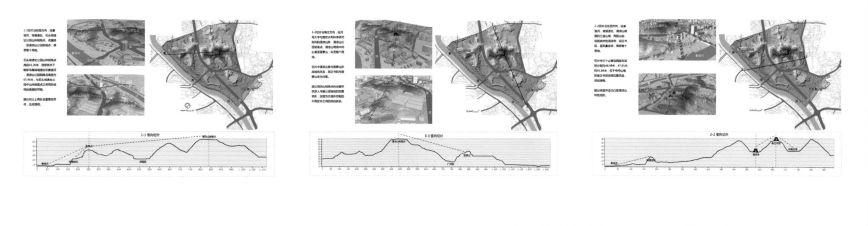

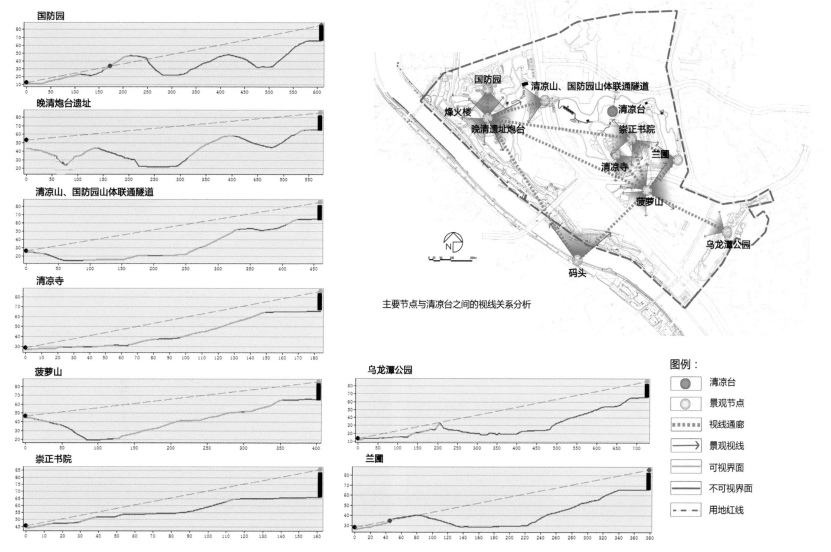

主要节点与清凉台之间的视线关系分析

图例：
- 清凉台
- 景观节点
- 视线通廊
- 景观视线
- 可视界面
- 不可视界面
- 用地红线

1	
2	3

1. 石头城遗址公园整体鸟瞰效果图1
2. 石头城遗址公园整体鸟瞰效果图2
3. 石头城遗址公园夜景鸟瞰效果图

1. 清凉台仰视效果图
2. 清凉台侧立面图
3. 清凉台正立面图
4. 清凉台效果图

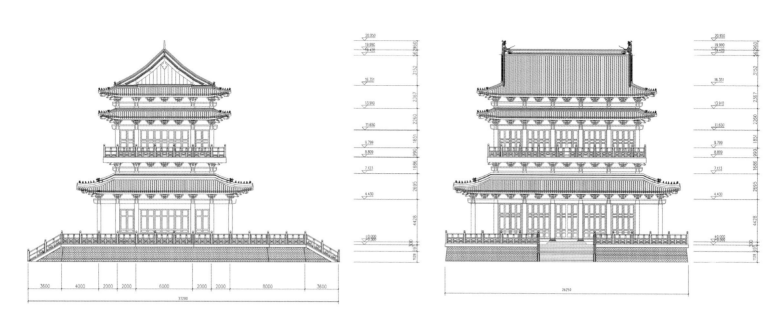

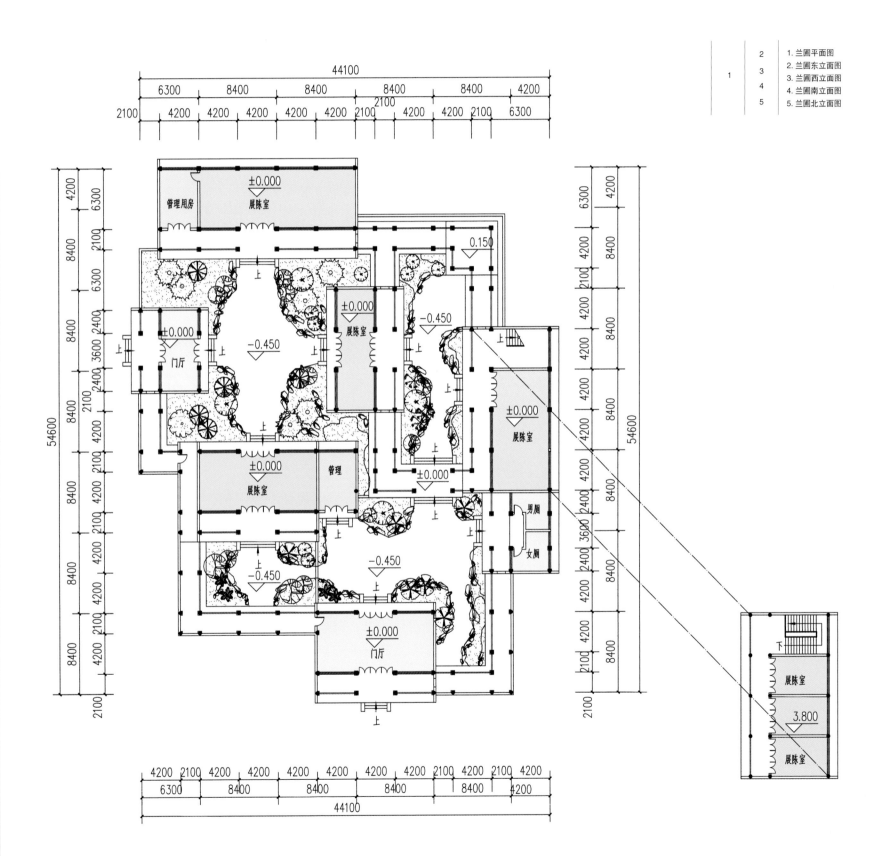

1. 兰圃平面图
2. 兰圃东立面图
3. 兰圃西立面图
4. 兰圃南立面图
5. 兰圃北立面图

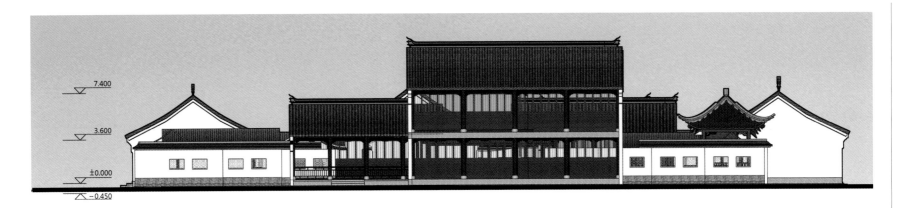
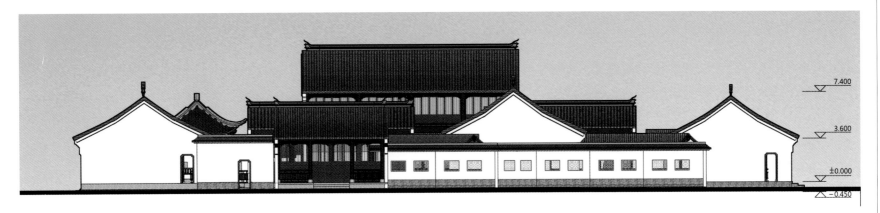
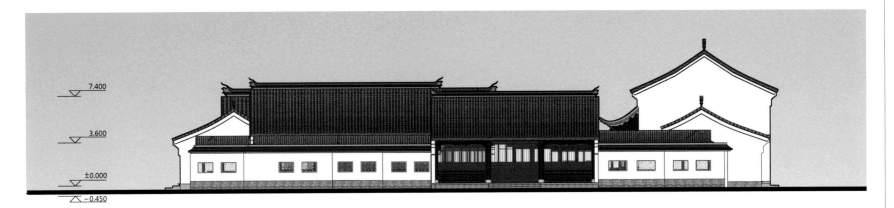
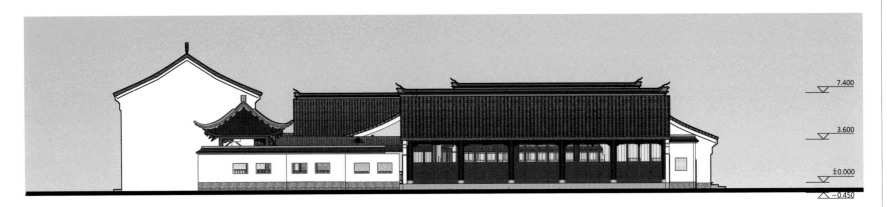

1. 兰圃鸟瞰图
2. 单坑遗址展示保护效果图
3. 坡面遗址坑展示保护效果图
4. 连续遗址坑展示保护效果图

城市公园 URBAN PARK | 129

大丰市五一湖商务公园设计
WUYI LAKE OFFICE PARK DESIGN, DAFENG

项目地点：江苏省大丰市	Location: Dafeng, Jiangsu	项目成员
设计时间：2012年	Design Period: 2012	景观设计：成玉宁、鲍洁敏、匡 纬
建成时间：2013年	Completion Time: 2013	建筑设计：成玉宁、李志刚、方 颖、汤春芳、陈 希、孙欣茹
委托单位：大丰市城市管理局	Client: Dafeng Urban Administration	结构设计：盛春陵、金 彦
用地面积：9392平方米	Area: 9392 m²	水电设计：李滨海、王晓晨

项目概况：

五一湖商务公园位于大丰市高新技术区，占地面积约 500 亩，是大丰市重点打造的花园城市办公休闲产业组团。公园作为高新区的"绿肺"，兼具产品展示、休闲旅游、健康产业等功能，彰显了大丰精致时尚、宜居的城市魅力。

设计策略

五一湖商务公园与五一河风光带、五一景观大道、园区创业服务中心等绿化景观工程，共同组成了高新区西片区核心，设计侧重处理公园与周边地块的关联。一方面，结合原有水系，通过适当挖填，形成自然水岸线；另一方面，通过微地形的处理，以凸显空间变化并延续、彰显场所肌理。设计沿湖布置造型新颖的展陈、休憩建筑，以呼应高新技术区的特质，满足园区商务及休闲活动需要。

Project Profile

Wuyi Lake Office Park is located in Dafeng High-tech Zone, covering an area of 500 mu. This is a key construction project of Dafeng Municipal Government, aiming to be a multinational site for both office property and recreation industry in the garden city. As the "green lung" of the hi-tech zone, this multifunctional park can meet the needs of product exhibition, recreation tourism and health industry. The design manifests that Dafeng, the exquisite and fashionable city, is suitable to live and full of charm.

Design Strategy

The core of the hi-tech zone's western section consists of Wuyi Lake Office Park and other green landscape projects, such as Wuyi River Scenic Belt, Wuyi Landscape Avenue, and the Business Incubator Center within the park. The design focuses on dealing with the connection between the park and its surroundings. On the one hand, the design integrates the original water system to form a natural shoreline through proper excavation and filling. On the other hand, through dealing with the microtopography, spatial changes are highlighted and the site's texture is maintained. Architectures of innovative design are arranged alongside the lake for rest and exhibition, which manifest the properties peculiar to the hi-tech zone, and also meet the demands of business and recreation activities.

C地块C₁建筑

城市公园 | URBAN PARK | 131

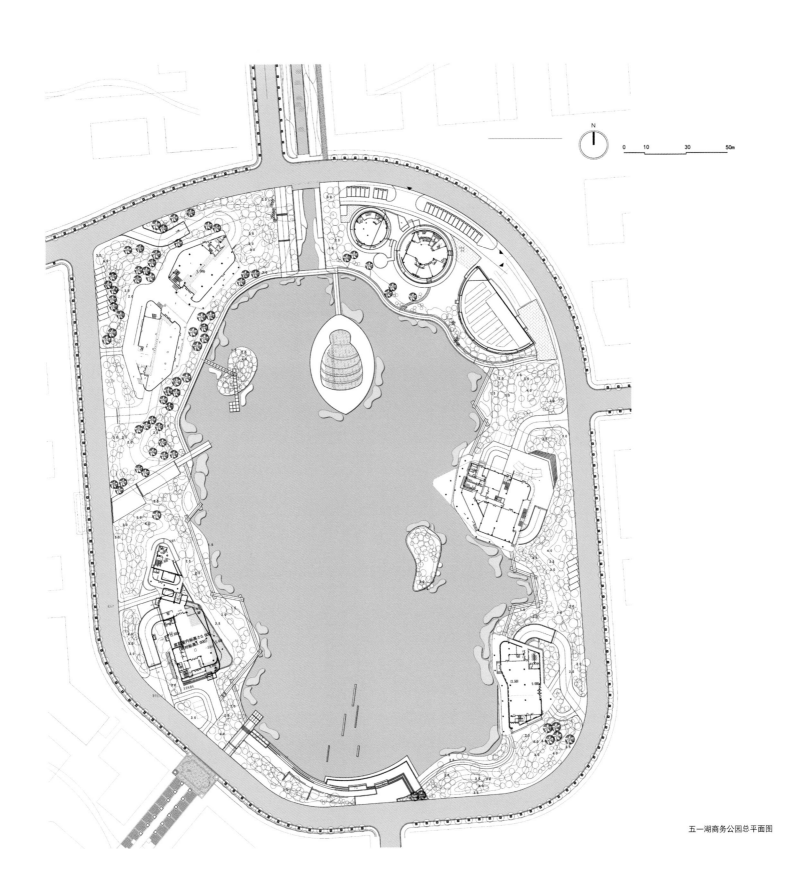

五一湖商务公园总平面图

城市公园 | URBAN PARK | 133

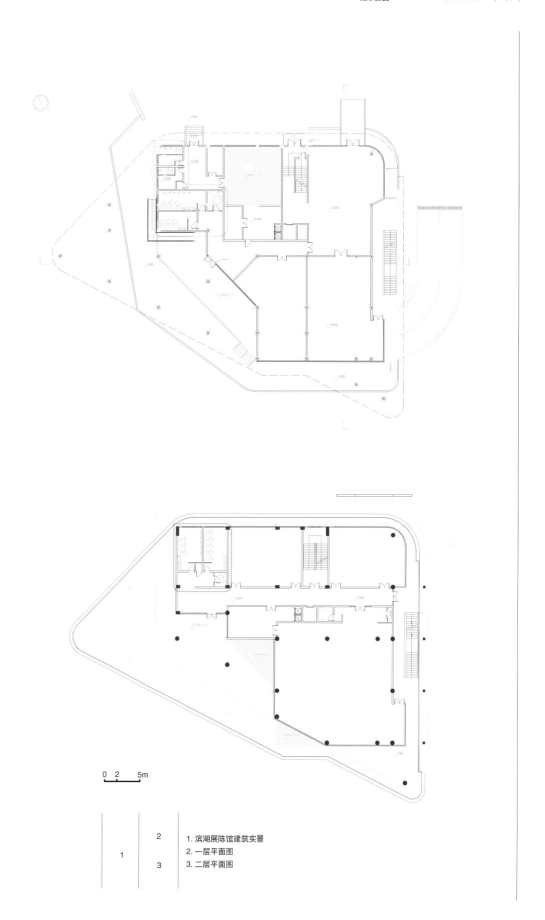

1. 滨湖展陈馆建筑实景
2. 一层平面图
3. 二层平面图

1. 展陈馆建筑实景
2. 展陈馆建筑立面图
3. 展陈馆建筑夜景

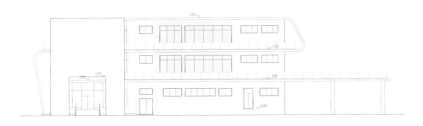
北立面 1:100

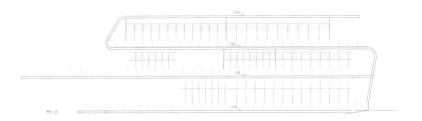
南立面 1:100

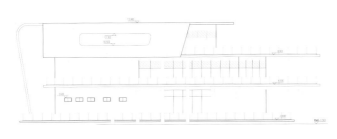
西立面 1:100

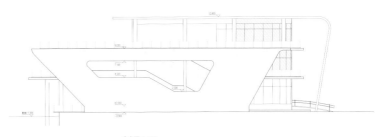
东立面 1:100

| 136 | URBAN PARK 城市公园

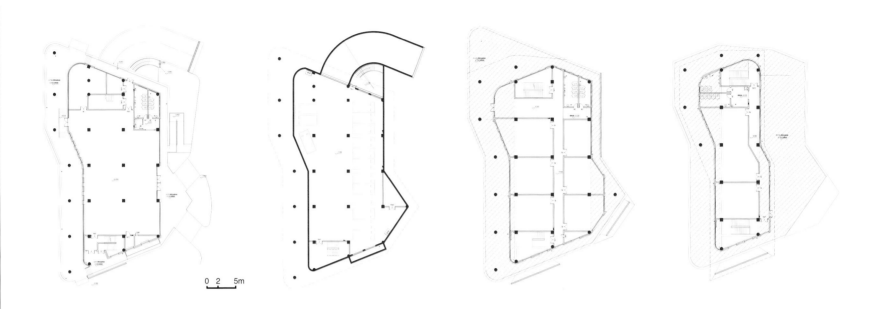

城市公园　URBAN PARK | 137

1	3
2	4	5

1. 展陈建筑各层平面图
2. 滨湖展陈建筑实景
3. 室内实景
4. 建筑剖面图
5. 建筑细部

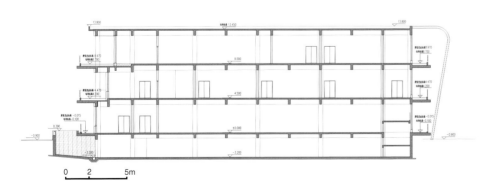

| 138 | URBAN PARK   城市公园

1. 滨湖展陈建筑夜景
2. 建筑立面图
3. 滨湖酒店建筑夜景

城市公园　URBAN PARK | 139

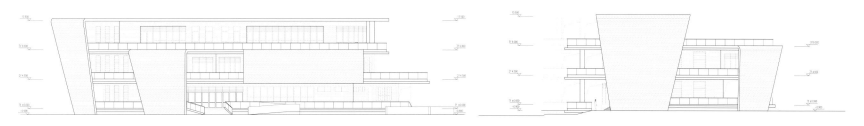

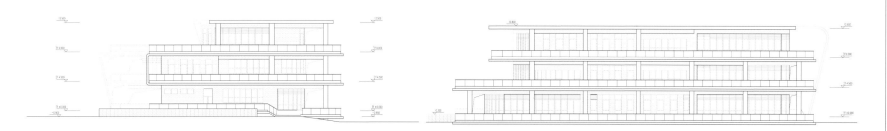

0　2　5m

城市公园　URBAN PARK | 141

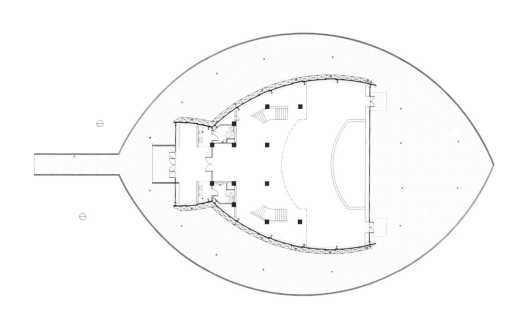
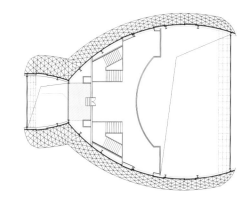
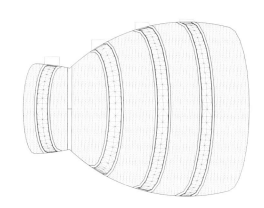

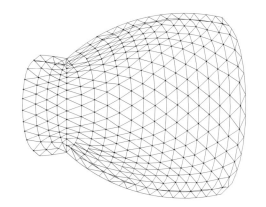

	2	1. 滨湖会所建筑实景
1	3　4	2. 一层平面图
	5	3. 二层平面图
		4. 屋顶平面图
		5. 屋顶桁架图

城市公园 | URBAN PARK | 143

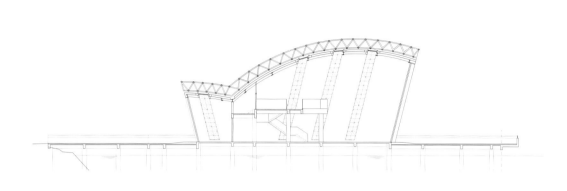

1	2	1. 滨湖会所建筑夜景
		2. 剖立面图
	3 4	3. 室内实景一
		4. 室内实景二

大丰市银杏湖公园设计
GINKGO LAKE PARK DESIGN, DAFENG

项 目 地 点：江苏省盐城大丰市	Location:	Dafeng, Jiangsu
设 计 时 间：2008 年	Design Period:	2008
建 成 时 间：2009 年	Completion Time:	2009
委 托 单 位：大丰市城市管理局	Client:	Dafeng Urban Administration
用 地 面 积：165600 平方米	Area:	165600 m²

项 目 成 员
景 观 设 计：成玉宁、王 玉、张 祎、房婉莹
建 筑 设 计：成玉宁、张亚伟、王天晖、戴丹骅
结 构 设 计：王伟成
水 电 设 计：刘 俊、王晓晨

项目概况

银杏湖公园作为大丰市南扩的一部分占地约 16 公顷。基地内部原有水面约占用地面积的 1/3，水体西侧为狭长的银杏林带。设计在对古典园林山水格局充分解读的基础上，结合场所的特点，营造出具有现代公园气息，又传承了中国园林意境的城市公园。

设计策略

1）传承中国园林经典空间意象

设计以"向传统致敬"为构思，解读并重构中国园林空间意向，融合场地肌理，塑造出以水面为核心的园林空间。

2）典型情境的再生

将原有场所特点通过抽象、再现手法作为公园景观节点，例如以"盘铁"表现传统制盐的工艺、用密植的灌木形成"麦浪"、以悬浮的表达绽放的"棉花"、以随风而动的光纤象征摇曳的"芦苇"……通过隐喻延续场所记忆。

Project Profile

Ginkgo Lake Park is part of Dafeng city's southward expansion project, covering an area of about 16 hectares. Within the site, the original water surface accounted for one third of the total area, with a stripe of Ginkgo forest belt standing to the east of the water. Based on a fully understanding of the classic garden landscape pattern, this design integrates the site's characteristics into the construction of a modern park which also carries the artistic conception of Chinese garden.

Design Strategy

1) Inherit the classic spatial image of Chinese garden

Inspired by the idea of "paying a tribute to tradition", this design interprets and restructures the spatial image of Chinese garden. Through integrating the site's texture, the design creates a garden space that focuses on water.

2) Regenerate the typical scenario

The design utilizes abstraction and reappearance to make the original site features as the park's landscape nudes. For example, "Pantie" (literally an iron pan used to boil salt) is set to display the traditional technique of making salt; bushes are planted compactly to form the image of "rippling wheat"; the suspending spray are used to express the image of "cotton blossom", fibers swaying with the wind is a symbol for waving "reeds"…In this way, metaphor is used to sustain the site's memory.

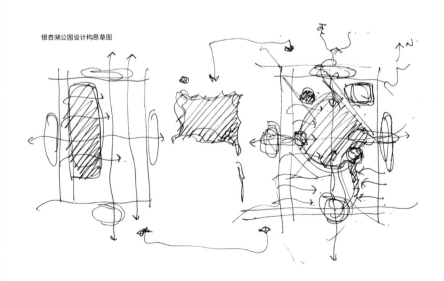

银杏湖公园设计构思草图

城市公园 URBAN PARK | 145

银杏湖公园总平面图

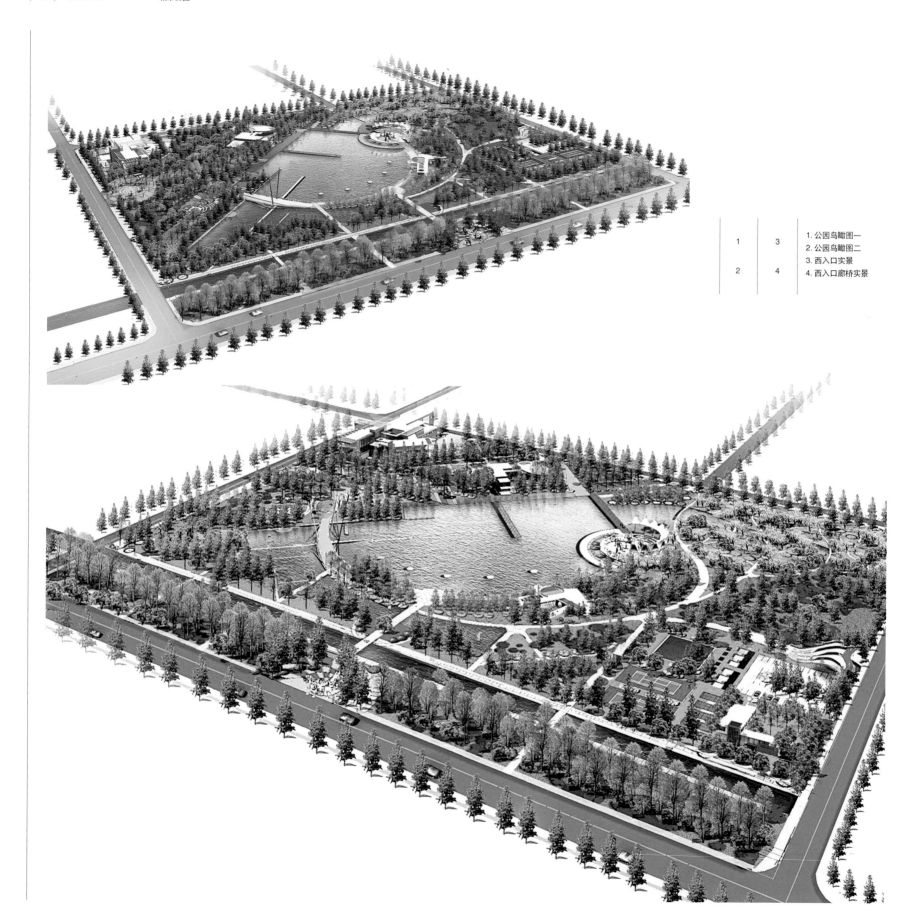

1. 公园鸟瞰图一
2. 公园鸟瞰图二
3. 西入口实景
4. 西入口廊桥实景

1. "蛙声十里"景观
2. 钢构平拱桥实景
3. 园路实景

餐饮休闲中心实景

城市公园　URBAN PARK | 153

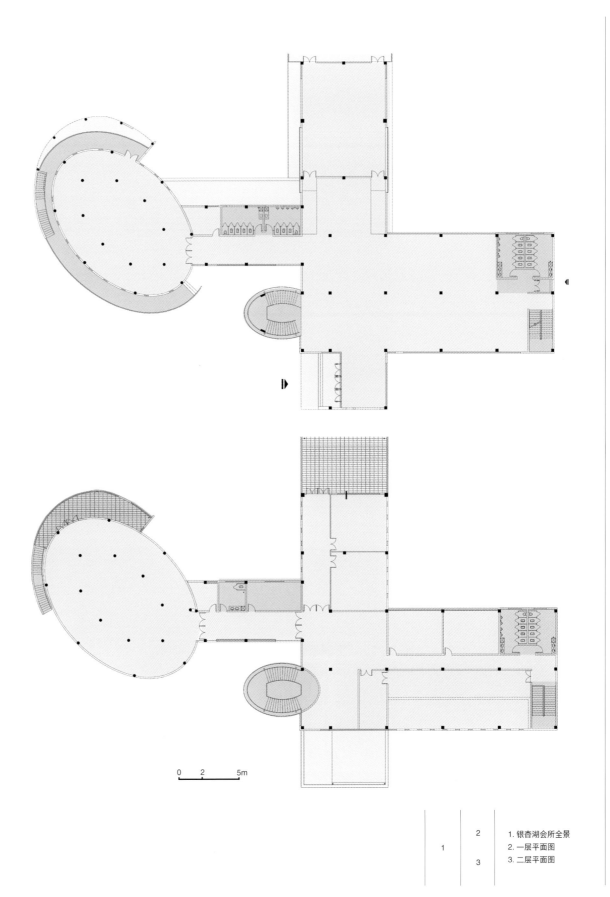

1. 银杏湖会所全景
2. 一层平面图
3. 二层平面图

| 154 | URBAN PARK   城市公园

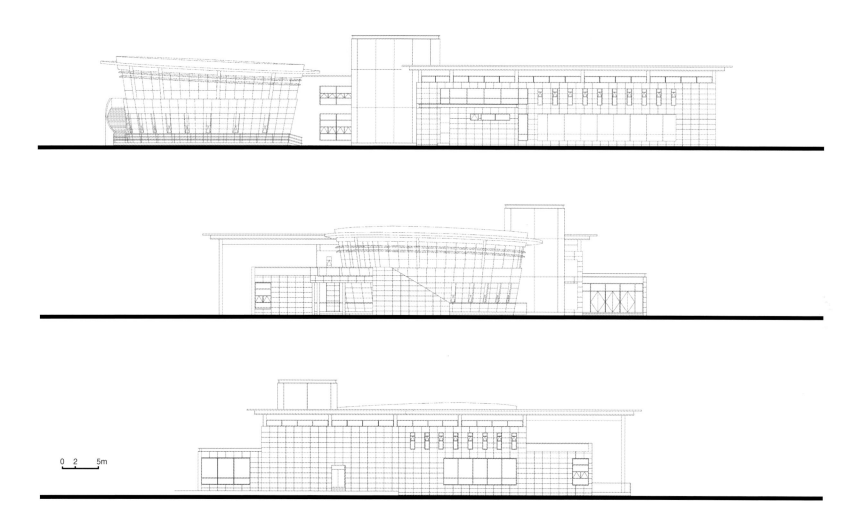

1. 银杏湖会所立面图
2. 银杏湖会所滨湖实景
3. 公园东入口广场实景

| 156 | URBAN PARK 城市公园

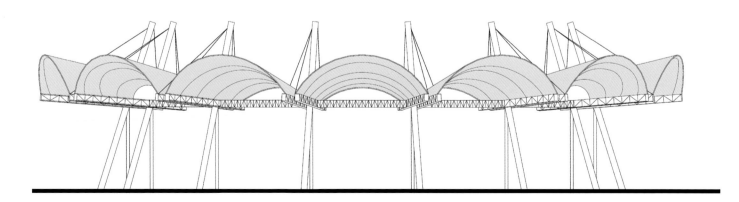

1	4	1. 小剧场平面图
2	3	2. 看台立面图
		3. 小剧场实景
		4. 小剧场夜景

公园钢构斜拉桥实景

淮安市中洲公园设计
ZHONGZHOU PARK DESIGN, HUAI'AN

项目地点：江苏省淮安市	Location: Huai'an, Jiangsu	项目成员
设计时间：2003 年	Design Period: 2003	景观设计：成玉宁、汪松陵
建成时间：2004 年	Completion Time: 2004	建筑设计：成玉宁、李 娜、李 勤
委托单位：淮安水利局	Client: Water Conservancy Bureau, Huai'an	结构设计：薛 峰
用地面积：24122 平方米	Area: 24122 m²	水电设计：刘 俊、王晓晨

项目概况

中洲公园位于淮安市古运河中段一个面积近 40 亩的岛上，是"古运河文化长廊"的一个重要景观节点。设计重现了淮安地标——清江浦楼，增建了文化中心，营造出饱含历史记忆又富有时代气息的新地标。

设计策略

设计充分研究环境空间环境，着重协调新建公园与周边城市空间尺度，以形成良好的对话关系。

1）空间形态的重组

设计通过适当增加竖向变化，丰富了空间层次，塑造出"山岛"的地貌特征，以表达"若隐若现，虚实相生"的趣味。中部的文化中心由有三组庭院组合而成，轴线灵活、空间多变，掩映于优美的季相植物景观之中。重建"清江浦楼"，增强了公园与城市及古运河景观轴线的对位关系。清江浦楼突兀水际，成为全岛及周边城市空间的视觉焦点。

2）地方建筑风格的传承与创新

清江浦楼及文化中心的设计融合了淮安地方建筑的传统形式与新技术条件，在满足现代审美及功能的同时，提炼、整合了场所原有的历史信息与淮安的传统建筑特色，并赋予其现代的空间表达形式，发展了淮安地区特有的建筑及景观风貌。

Project Profile

Located in an island of 40 mu which lies in the middle part of the ancient canal in Huai'an Zhongzhou Park is an important landscape node of the "Cultural Corridor of Ancient Canal". The design reconstructs Qingjiangpu Tower, the landmark of Huai'an, and builds out a cultural center, which makes the new landmark filled with not only historic memories but also modern senses.

Design Strategy

With the spatial environment completed studied, the design put emphasis on coordinating the spatial scale between the newly built park and its surrounding cities so as to establish good dialog relationship.

1) Spatial form reorganization

The design enriches the spatial level through appropriately adding vertical changes. It brings in the landscape characteristics of "hills and islands" to express the delight of "shifting from visible scenes to hidden scenes, integrating virtual images into true ones". The cultural center, Located in the central section, consists of three courtyard compositions. With flexible axes and changeable spaces, the center stands among the beautiful seasonal plant landscape. The reconstruction of Qingjiangpu Tower strengthens the para-position relationship between landscape axes of the park, the city and the ancient canal. Qiangjiangpu Tower stands out over the water, becoming a visual focus of the whole island and the neighboring cities.

2) The inheritance and innovation of regional architectural style

The design of Qingjiangpu Tower and the culture center integrates the traditional form and new-tech conditions of regional architectures in Huai'an. With modern spatial expression added, the design not only meets modern aesthetic demands, but also develops distinctive architecture and landscape peculiar to Huai'an through extracting and blending the original historical information as well as Huai'an traditional architectural features.

城市公园 | URBAN PARK | 161

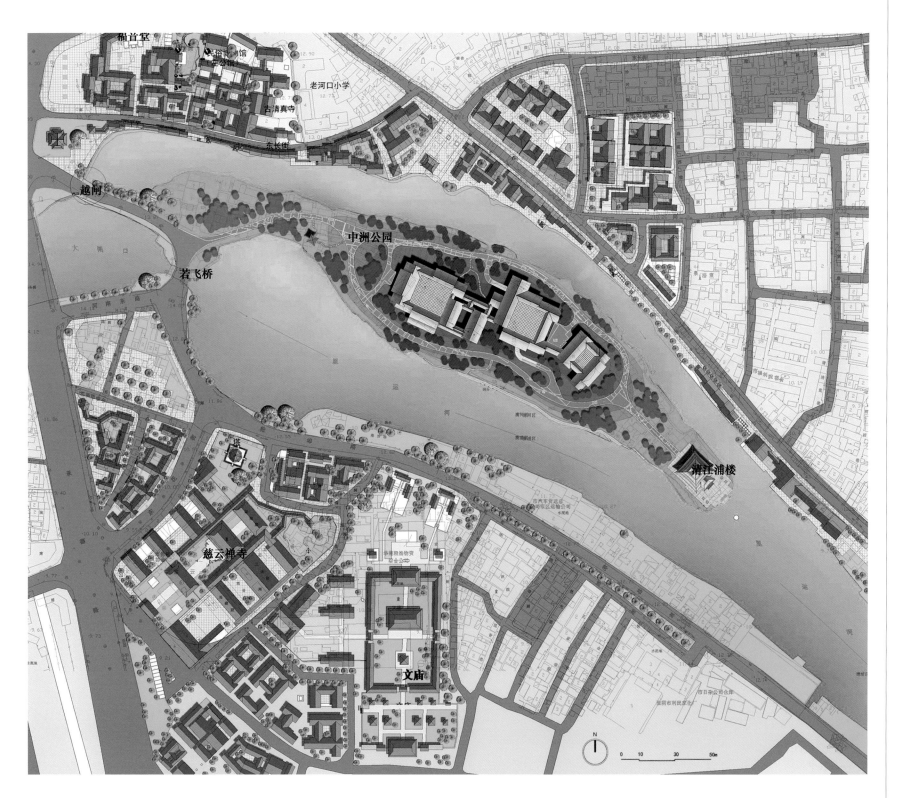

中洲公园总平面图

1. 中洲岛鸟瞰实景
2. 公园主园路实景

城市公园　URBAN PARK　| 163 |

URBAN PARK 城市公园

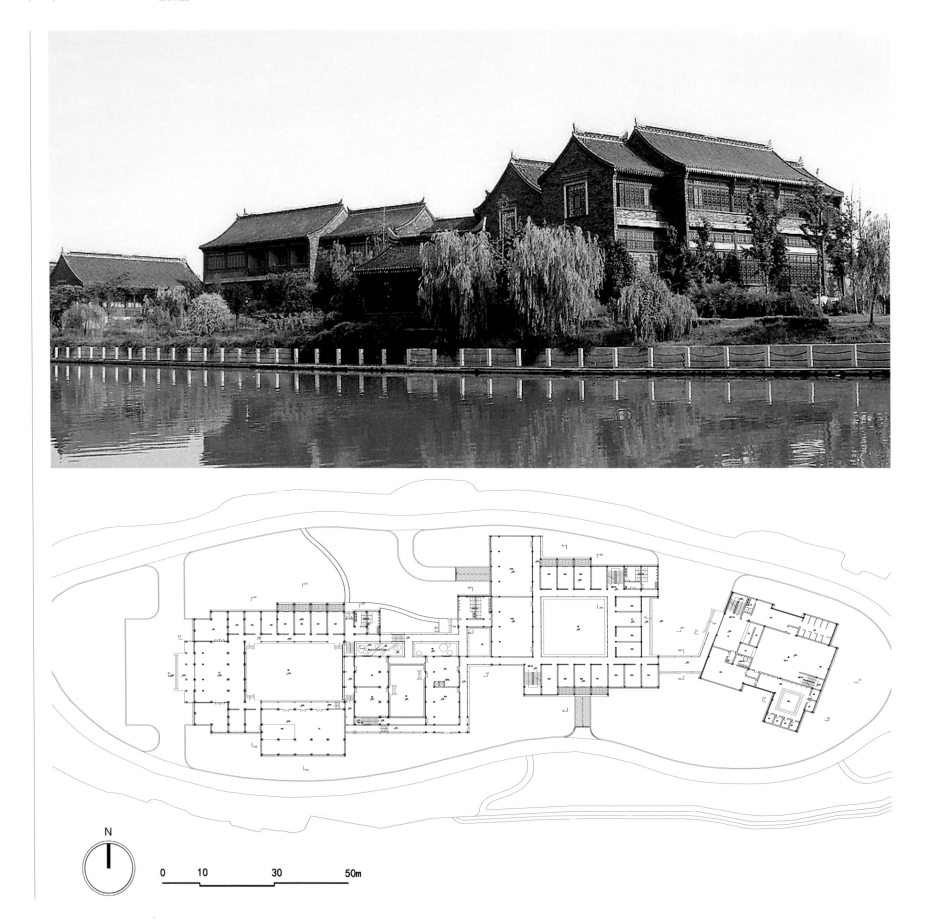

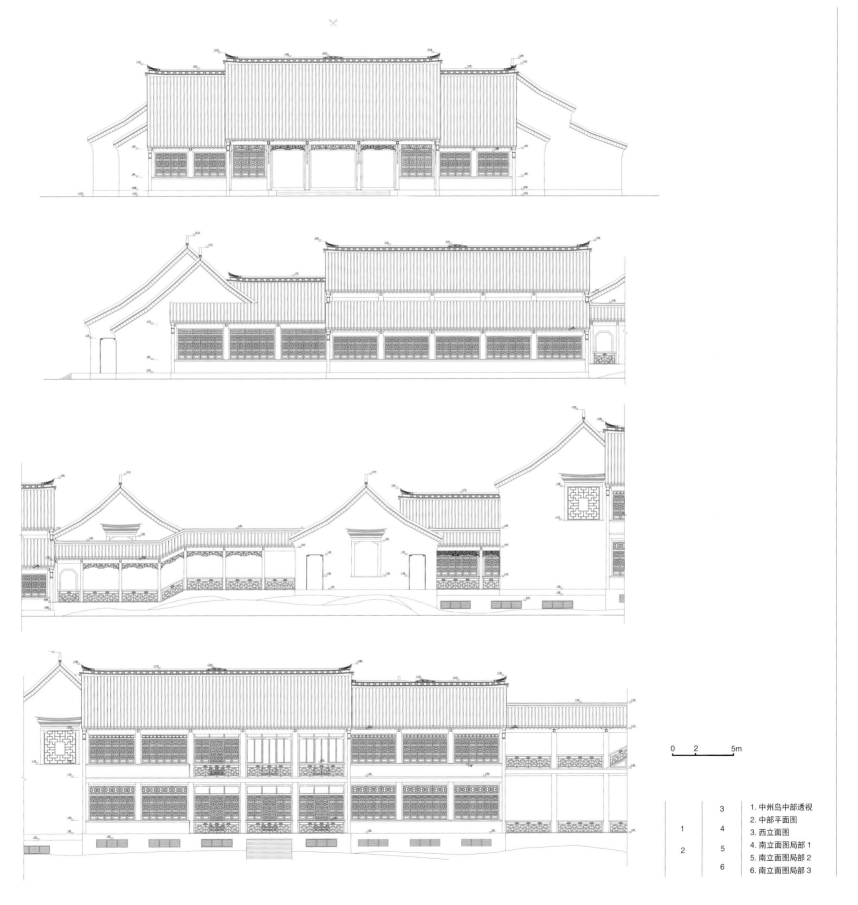

1. 中州岛中部透视
2. 中部平面图
3. 西立面图
4. 南立面图局部1
5. 南立面图局部2
6. 南立面图局部3

1. 淮安名人馆东北透视
2. 淮安名人馆西北透视
3. 清江浦楼实景

1. 园路种植实景
2. 夕照亭实景
3. 建筑山墙细部

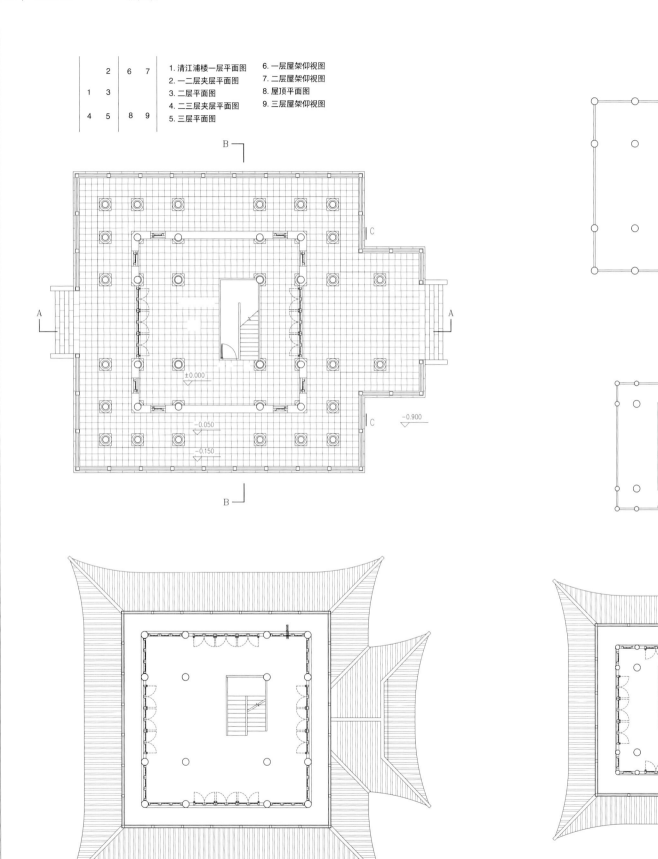

	2	6	7
1	3		
4	5	8	9

1. 清江浦楼一层平面图
2. 一二层夹层平面图
3. 二层平面图
4. 二三层夹层平面图
5. 三层平面图
6. 一层屋架仰视图
7. 二层屋架仰视图
8. 屋顶平面图
9. 三层屋架仰视图

城市公园 | URBAN PARK | 171

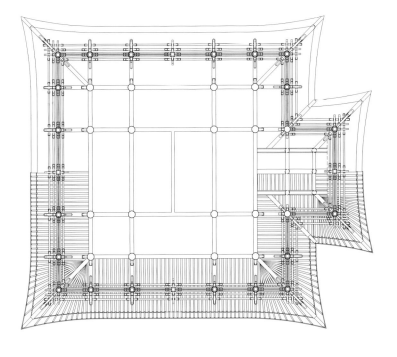
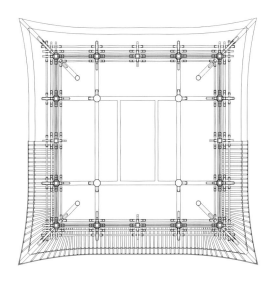
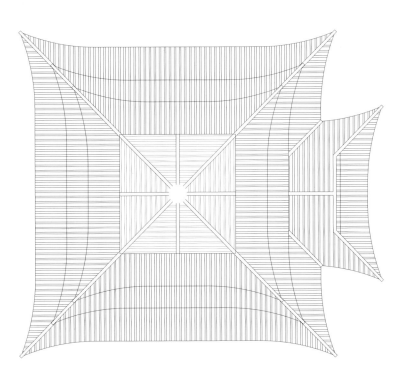
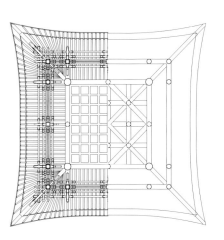

1. 清江浦楼全景
2. 正立面图
3. 侧立面图
4. B-B 剖面图
5. A-A 剖面图

城市公园　URBAN PARK | 173

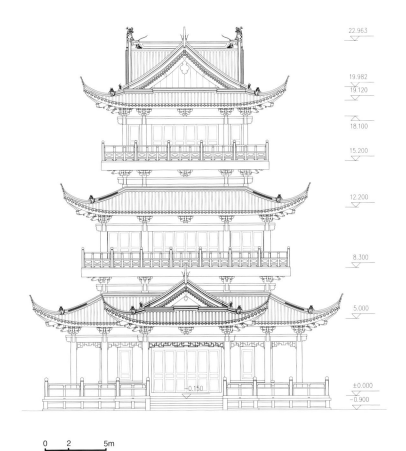
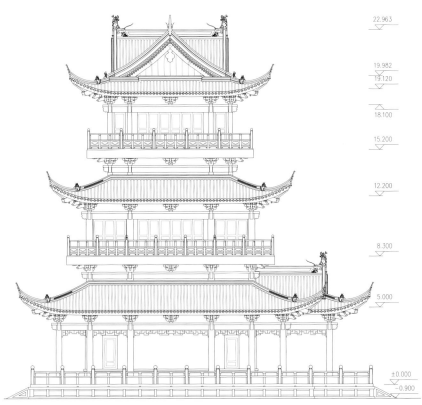
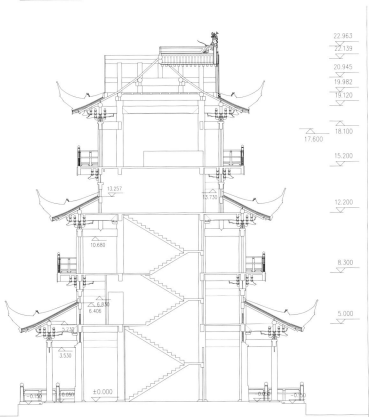
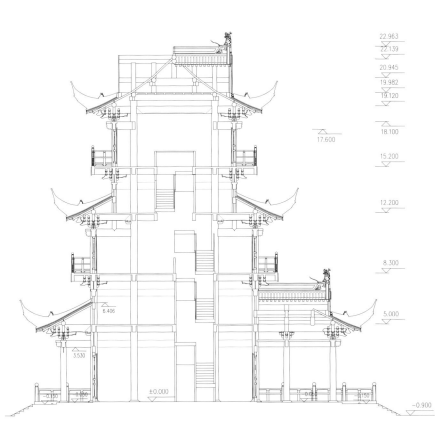

扬中市滨江公园设计
YANGTZE RIVERSIDE LANDSCAPE DESIGN, YANGZHONG

项目地点：江苏省扬中市	Location: Yangzhong, Jiangsu	项目成员
设计时间：2010年	Design Period: 2010	景观设计：成玉宁、袁旸洋、王巍巍、苏雅茜
建成时间：2012年	Completion Time: 2012	建筑设计：成玉宁、袁旸洋、王巍巍、崔巍
委托单位：扬中市市政园林局	Client: Bureau of Urban Utilities and Landscaping of Yangzhong	结构设计：王伟成、马军、赵才其
用地面积：216379.94平方米	Area: 216379.94 m²	水电设计：李滨海、王晓晨

项目概况

滨江公园位于扬中市区南部，毗邻长江，用地呈狭长的带状，占地面积约21公顷。由于原工业堆场、沙石码头等人工设施的长期扰动，造成了滨江岸线生态条件的恶化。设计致力于恢复和优化沿江生态岸线和滨水环境，满足现代高品质的滨江生活要求，展现了滨江地段的生态与人文魅力。

设计策略

遵循生态修复的方法，设计对场所环境进行有序的引导性修复，满足人们欣赏江滨、回归自然的愿望。同时，选择性地保留与彰显了场所记忆，赋予滨江岸线新的活力。

1) 人工引导下的生态修复

针对长江岸线受到侵蚀、破坏的状况，设计采取抛毛石并间种植物的方法，既有效应对江水的潮汐变化，又创造了多孔空间，为沼生动植物提供了生存环境。从北部的城市到南部的长江，设计采用"梯度"策略营造空间及生境，实现了由人工向自然的过渡。

2) 地带性景观特征的营造

设计中大量运用落羽杉、垂柳、杞柳、芦竹、芦苇、荻草、蒲苇、芒草等地带性适生植物。同时，丰富江堤景观，结合原江堤加固工程营建曲线型台地并种植亚热带适生湿地植物150余种，形成了地带性湿地植物品种园区。

3) 可再生材料的运用

景观的营造坚持可持续的理念，大量使用可回收的金属材料。凡是存在有季节性淹没可能的栈道等设施，其建造材料均采用不饰涂装的拉丝不锈钢材质，从而实现免维护。

4) 场所记忆的响应与表达

园内的建、构筑物的设计力求简洁、明快。"迎江阁"的构思取意"春江水暖鸭先知"；游船码头的创作灵感源于"鹤立江滩"，以唤起人们对旧时船坞、吊车的记忆；作为湿地科普馆的"望江台"状如螺壳，极具形式感且满足了科普展示的空间及流线需求。

Project Profile

Adjacent to the Yangtze River, Yangtze Riverside Park is located in the south of Yangzhong. With a total area of nearly 21 hectares, the landform of this site appears as a long and narrow ribbon. Due to the long-term disturbance from artificial facilities like industrial storage yard and dock, the Yangtze riverside ecological condition suffered a lot from deterioration. This design aims to restore and optimize the riverside eco-environment, and meet the demands of high quality riverside living, thus presenting the ecological and cultural charm of the riverside region.

Design Strategy

This design adopts the approaches of ecological renovation. To meet people' desire for appreciating riverside scenes and returning to the nature, the site's environment is renovated by sequential guidance. Meanwhile, the site memories are selectively maintained, which injects new energy into the shoreline of Yangtze River.

1) Ecological renovation under human guidance

To handle the erosion and destruction of the shoreline of Yangtze River, ashlars are poured and plants are interplanted. This method can not only effectively confront the tidal variation, but also creates porous spaces which are home to paludose animals and plants. From the northern city to the southern Yangtze River, the design adopts "gradient" strategy to construct spaces and habitats, which realizes the transition from being artificial to natural.

2) The creation of zonal landscape features

This design utilizes a large amount of viable zonal plants, including white cypress, willow, Salix, reed bamboo, reeds, silver grass, pampas grass, and Miscanthus. Meanwhile, the riverbank landscape is enriched. The original riverbank reinforcement is blended into the construction of a arc platform where plant over 150 species of wetland plants that are suitable to living in tropical regions. In this way, a park for zonal wetland plant species is established.

3) The application of renewable materials

To practice the idea of sustainable development, a large number of recycled metal materials are used for the construction. Since facilities, such as plank roads alongside cliffs, may be possible to suffer from seasonal floods, bushed stainless steel without coating is adopted as the corresponding construction materials to obtain the goal of maintenance-free.

4) The response to and the expression of site memories

The construction is designed to be brief and bright. The construction of Yingjiang Pavilion is inspired by an ancient poem,"the duck knows first when the river becomes warm in spring". And the construction idea of the marina is derived from a four character Chinese phrase "Helijiangtan" (literally a crane stands on the river beach). These are used to remind people of docks and hoists in the old times. The "Wangjiang Terrace" of the Wetland Science Museum is shaped like a conch shell, which not only shows the beauty of streamline but also provides enough space for scientific exhibition.

扬中滨江公园设计构思草图

| 176 | URBAN PARK 城市公园

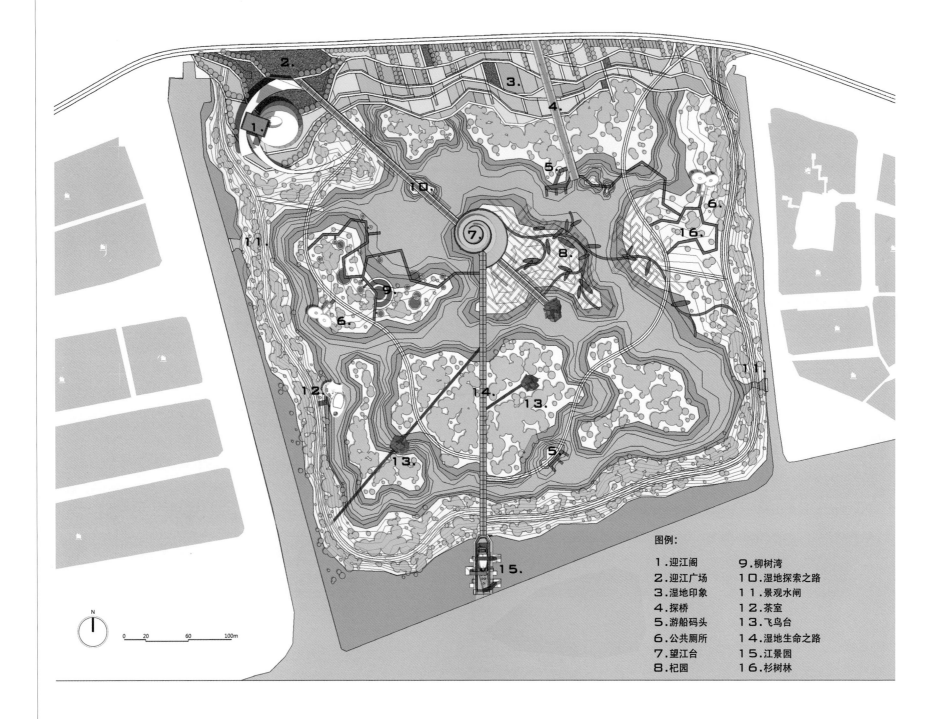

图例：

1. 迎江阁
2. 迎江广场
3. 湿地印象
4. 探桥
5. 游船码头
6. 公共厕所
7. 望江台
8. 杞园
9. 柳树湾
10. 湿地探索之路
11. 景观水闸
12. 茶室
13. 飞鸟台
14. 湿地生命之路
15. 江景园
16. 杉树林

1				1. 扬中滨江公园总平面图
	2	3	4	2. 木栈道及水生植被实景
				3. 木栈道实景
	5			4. 游艇码头实景
				5. 杞柳园

城市公园　URBAN PARK | 177

1. 公园鸟瞰实景
2. 入口广场景观
3. 廊柱花海实景
4. 入口花坛实景

城市公园 | URBAN PARK | 179

| 180 | URBAN PARK 城市公园

1. 迎江阁立面图
2. 幕墙细部实景
3. 一层面图
4. 二层面图
5. 迎江阁实景

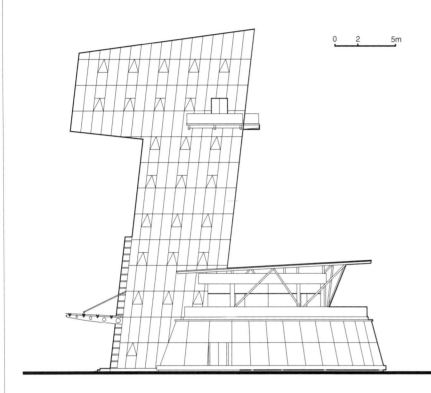

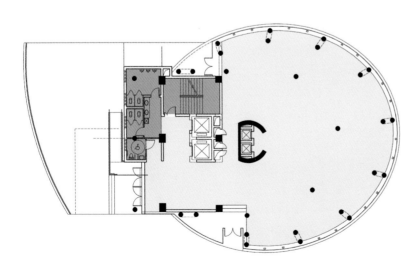
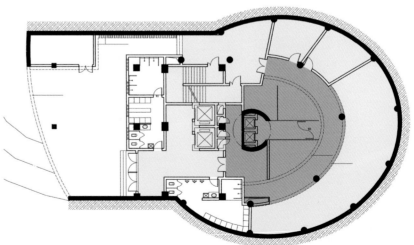

| 182 | URBAN PARK 　　城市公园

1	2	
3		4

1. 公园入口实景
2. 迎江阁仰视
3. 中部透视
4. 迎江阁石滩

| 184 | URBAN PARK 城市公园

城市公园　URBAN PARK | 185

1. 望江台及木栈道实景
2. 望江台内部实景
3. 望江台及钢栈桥实景
4. 一层平面图
5. 二层平面图
6. 剖面图

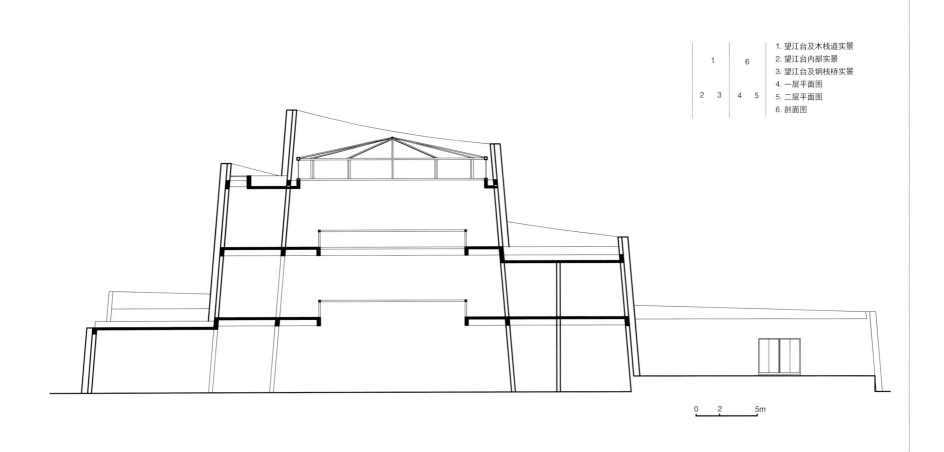

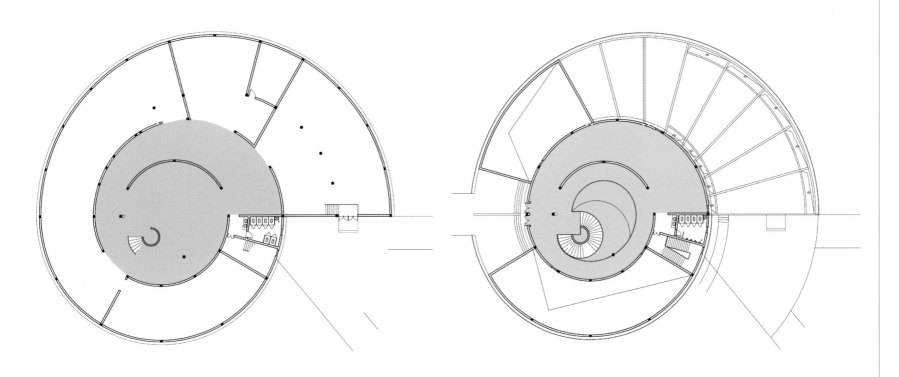

| 186 | URBAN PARK   城市公园

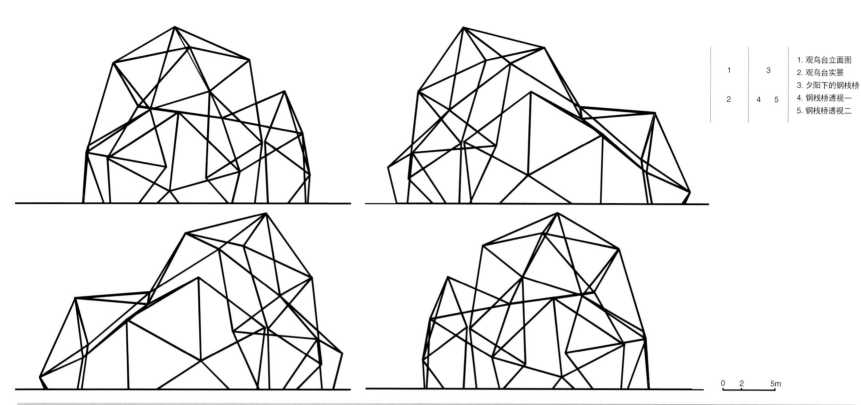

| 1 | 3 |
| 2 | 4 | 5 |

1. 观鸟台立面图
2. 观鸟台实景
3. 夕阳下的钢栈桥
4. 钢栈桥透视一
5. 钢栈桥透视二

芦苇丛中的木栈道与观鸟台

| 190 | URBAN PARK 城市公园

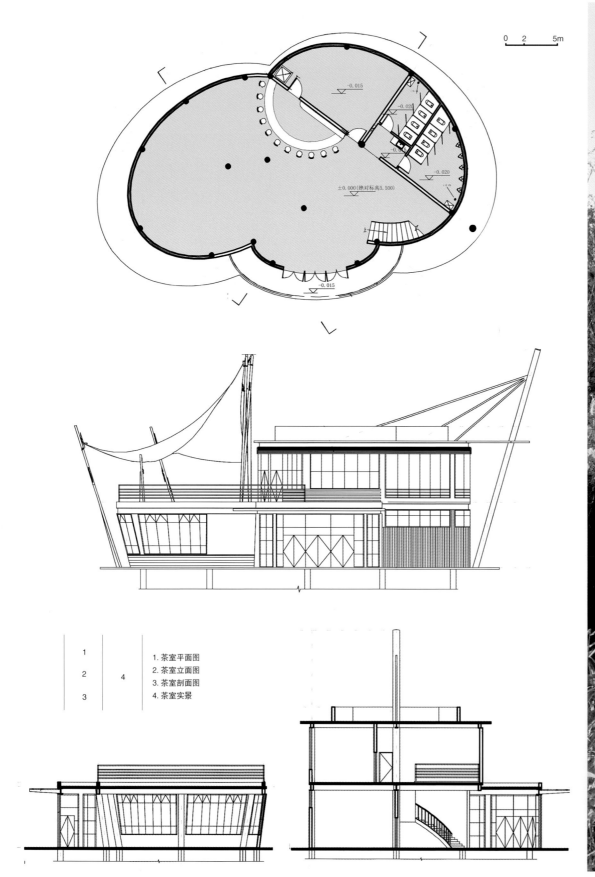

1. 茶室平面图
2. 茶室立面图
3. 茶室剖面图
4. 茶室实景

城市公园　URBAN PARK | 193

| 1 | 1. 远眺"鹤立江滩"码头 |
| 2 | 2. 钢栈桥实景 |

城市公园　　URBAN PARK

1. 湿地栈道实景
2. 杉林雾森实景
3. 暮色中的钢栈桥

| 1 | 2 | 1. 滨江栈桥实景
2. 滨江码头实景

滨水景观
WATERFRONT LANDSCAPE

宿迁市河滨公园设计
RIVERSIDE PARK DESIGN, SUQIAN

项目地点：江苏省宿迁市	Location: Suqian, Jiangsu	项目成员
设计时间：2007年	Design Period: 2007	景观设计：成玉宁
建成时间：2008年	Completion Time: 2008	建筑设计：成玉宁、王天晖
委托单位：宿迁市园林局	Client: Suqian Municipal Bureau of Parks	结构设计：王伟成
用地面积：3.6公顷	Area: 3.6 ha	水电设计：刘俊、刘向上

项目概况

河滨公园位于宿迁市黄河故道畔，为南北向的带状绿地，紧邻城市中心干道。场地原为沿河堆场等，设计着重恢复用地范围内的生态环境，营造了满足市民户外生活的滨水空间。

设计策略

在场所研究的基础上，于黄河一号、二号桥之间设置滨水栈道，打通黄河风光带的滨水观赏及游览线路。在带状空间中，使用斜向穿插的廊道，形成富有节奏与韵律的景观空间，在打破带状空间的单一性的同时，增强了城市与古黄河空间的联系。设计采用缓坡草地取代了原有的二级驳岸，以满足市民的亲水需求。

Project Profile

Riverside Park, a long and narrow north-south green belt, is located alongside the old course of Yellow River, and is adjacent to the artery of the central city. The site used to be riverside storage yards and other artificial facilities. This design focuses on the eco-environment renovation within the site and constructs a riverside space which can meet citizens' needs of outdoor activities.

Design Strategy

Based on site researches, the waterside plank road is set between No.1 and No.2 bridges of Yellow River, which opens up a waterside touring route of the Yellow River scenic belt. In the belt-shaped space, galleries are alternately arranged in a tilt direction, creating a landscape space with rhythm and rhyme. In this case, the unity of the belt-shaped space is broken, and at the same time, the spatial connection between the city and the ancient Yellow River is strengthened. To meet citizens' need of being closer to the water, meadows with gentle slopes are set in place of the original second-level revetment.

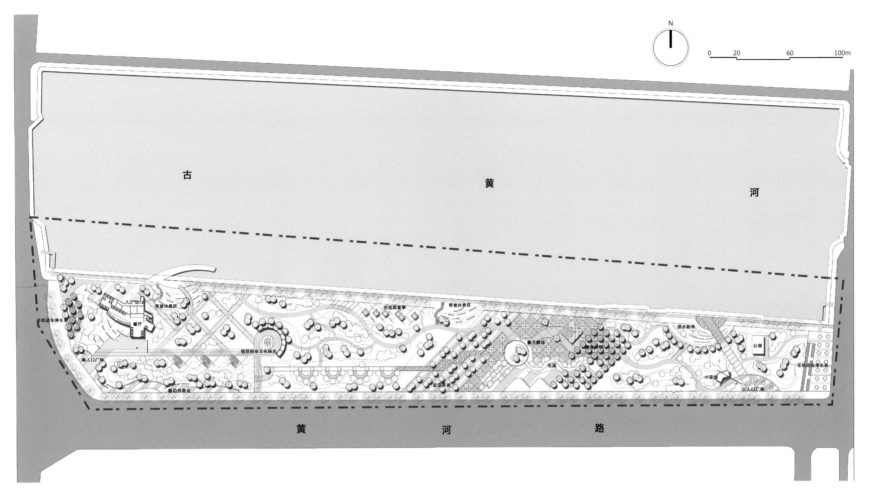

河滨公园总平面图

河滨公园鸟瞰图

公园看台实景

1. 滨河步道实景
2. 休闲广场与观景亭

滨水景观　WATERFRONT LANDSCAPE | 203

1. 坡地花园实景
2. 观景亭实景

| 206 | WATERFRONT LANDSCAPE 滨水景观

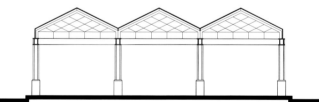

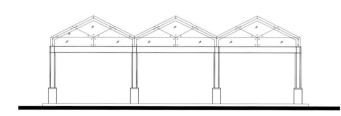

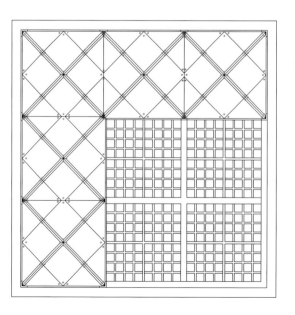

1		4	1. 休息亭实景
			2. 立面、剖面图
			3. 顶平面图
2	3	5	4. 园路实景一
			5. 园路实景二

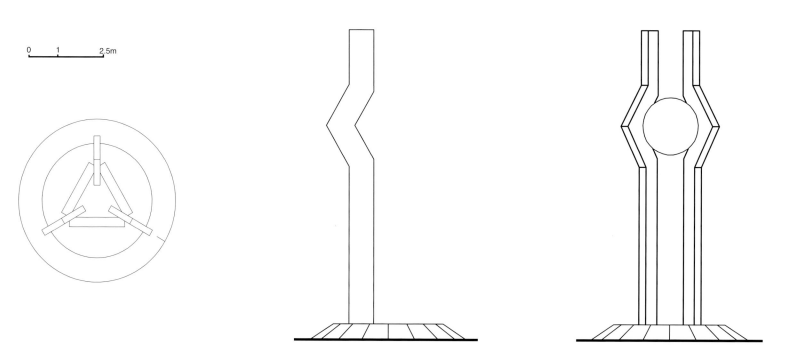

滨水景观 WATERFRONT LANDSCAPE | 209

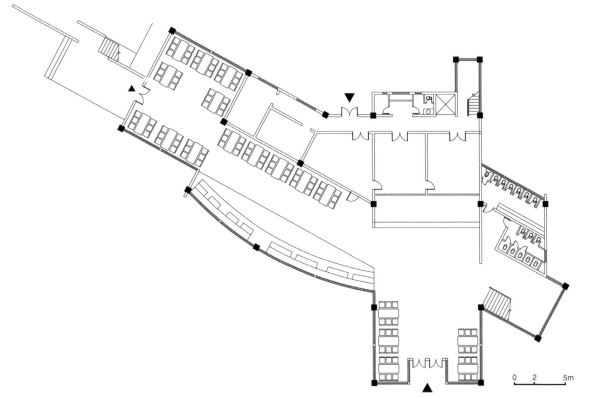

1	4
2 3	5

1. 创业园实景
2. 标志平面图
3. 标志立面图
4. 餐厅一层平面图
5. 餐厅实景

大丰市二卯酉河景观带设计
ERMAOYOU RIVERSIDE LANDSCAPE DESIGN, DAFENG

项目地点：江苏省大丰市	Location: Dafeng, Jiangsu	项目成员
设 计 时 间：2003 年	Design Period: 2003	景观设计：成玉宁、张 楷、丁 颖、李雯婷
建 成 时 间：2004 年	Completion Time: 2004	建筑设计：成玉宁、张 楷
委 托 单 位：大丰市城管局	Client: Dafeng Municipal Urban Management Bureau	结构设计：薛 峰
用 地 面 积：6.2 公顷	Area: 6.2 ha	水电设计：刘 俊、刘向上

二卯酉河景观带总平面图

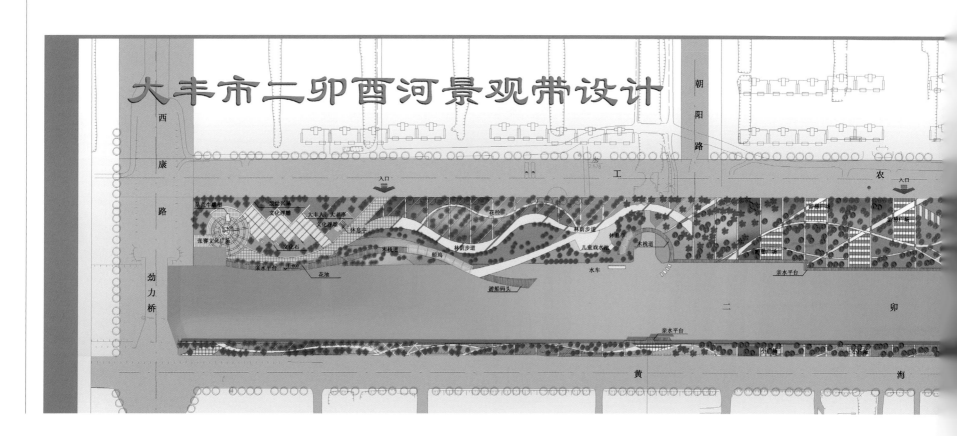

项目概况

二卯酉河景观带位于大丰市东西轴线的西端。人工开挖的二卯酉河在历史上曾发挥过灌溉农田、泄涝排洪及运输的重要作用，如今成为一处承载市民记忆的游憩、生活空间。

设计策略

1) 重组场地肌理

设计以原场所中大面积存在的南北向"沟渠"作为基本构图母题，凸显水网交错的场所肌理。通过南北视轴线的沟通，整合二卯酉河及两岸绿地。在研究视线透景的基础上，设置对景突出景观的可识别性。

2) 艺术地表达历史

设计围绕张謇"废灶兴垦"这一历史题材，于景观带西端打造张謇纪念园。采用浮雕、地雕艺术地再现民族实业家张謇的故事。

3) 再现典型景观

景观带中部设有"船"形茶餐厅，通过建筑形态隐喻二卯酉河在历史上所具有的航道功能及入海位置。结合人们休憩活动的需求，点缀水车、风车等地方性特色的生活场景，以呼唤场所记忆。

Project Profile

Ermaoyou Riverside Landscape is located at the eastern end of Dafeng's east-west axis. Ermaoyou River is a man-made river, which used to play an important role in farmland irrigation, flood drainage and transportation. By now, it has become a recreation and living space which carries a lot of memories of the citizens.

Design Strategy

1) Reorganize the site texture

The design uses the north-south "ditches" which cover a large area of the original site as its basic composition motif, and highlights the site texture of the interlaced water networks. Thorough the intersection of the northern and southern axes, the design integrates Ermaoyou River and the riverside green land. Based on the study of visual relationship, opposite scenery is set to stress the landscape identity.

2) The artistic express of history

Centered on the historical story that Zhang Qian discarded saltworks and enclosed tideland for cultivation, Zhang Qian Memorial Park is built at the western end of the landscape. This national industrialist's story is represented through the art of relief sculpture and floor sculpture.

3) The reappearance of typical landscape

A boat-shaped tea café is set in the landscape center, the metaphor of which implies the transportation role that Ermaoyou River played in the history. To cater for people's need of rest and activities, distinctive scenes of life, such as waterwheel and windmill, are dotted to recall memories of the site.

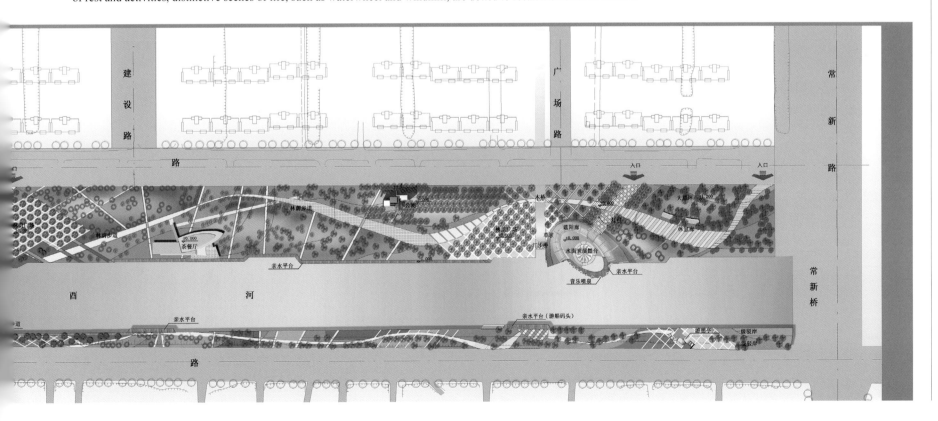

1. 健身乐园广场实景
2. 滨水码头实景
3. 张謇纪念园实景
4. 休憩亭周边实景

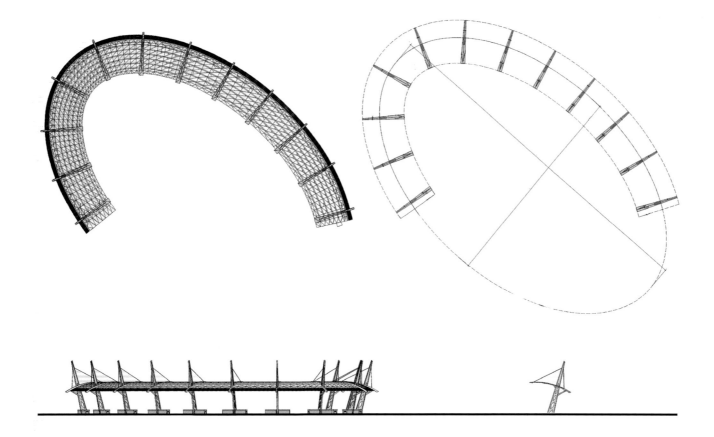

1. 小剧场看台屋顶与平面图
2. 小剧场立面、剖面图
3. 小剧场实景
4. 小剧场仰视

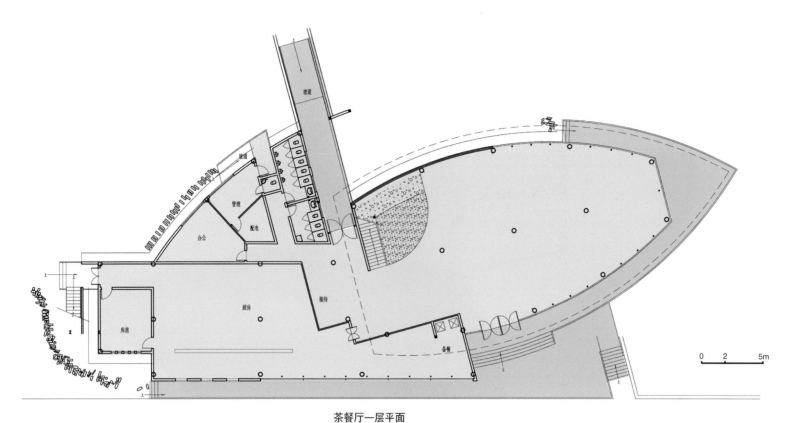

茶餐厅一层平面

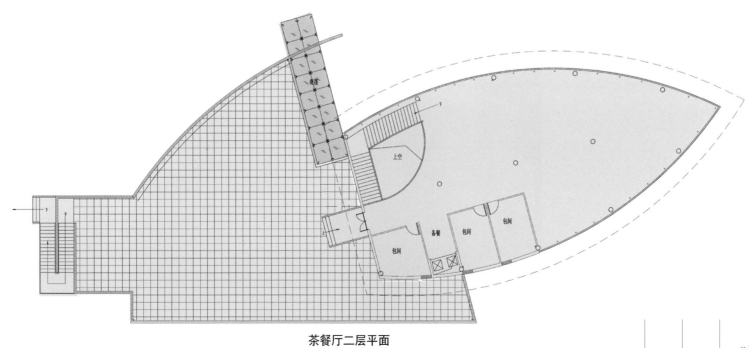

茶餐厅二层平面

1. 茶餐厅一、二层平面图
2. 茶餐厅实景

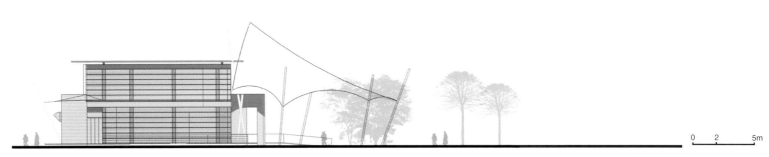

茶餐厅东立面

2		1. 茶餐厅外墙细部
1	4	2. 茶餐厅无障碍坡道
3		3. 茶餐厅透视
		4. 茶餐厅立面图

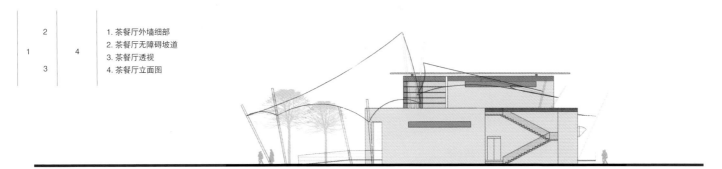

茶餐厅西立面

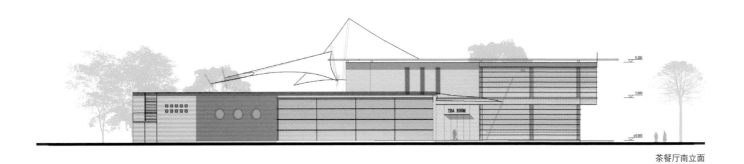

茶餐厅南立面

茶餐厅北立面

滨水景观　WATERFRONT LANDSCAPE | 221

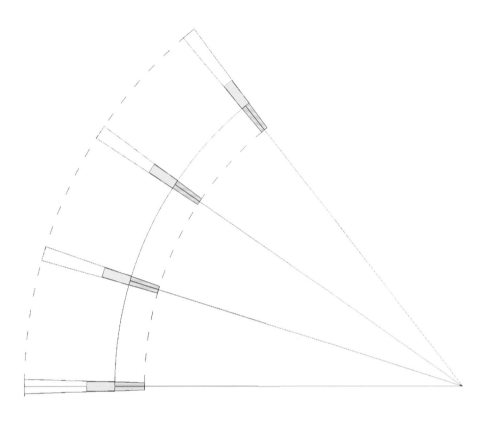

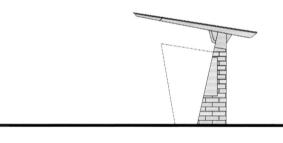

1			1. 水车亭实景
	2	3	2. 水车亭平面图
			3. 水车亭顶平面图
		5	4. 水车亭实景
	4	6	5. 水车亭剖面图
			6. 水车亭西立面图

0　0.5　1.5m

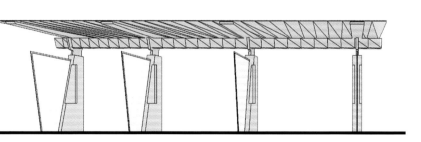

2		1. 休息廊实景
1 3 5		2. 休息廊顶平面图
		3. 休息廊立面图
4		4. 休息廊剖面图
		5. 休息亭实景

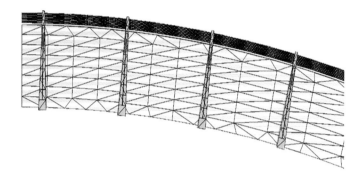

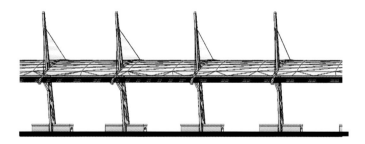

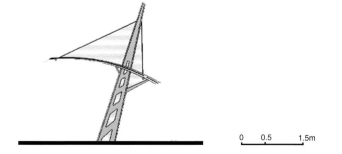

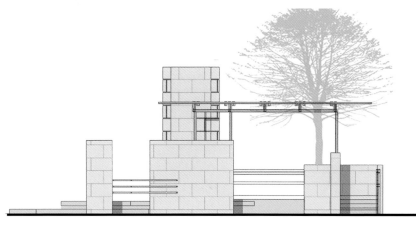

休息亭南立面

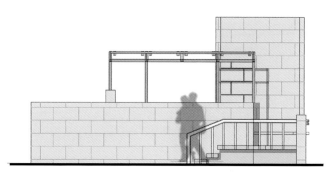

休息亭东立面

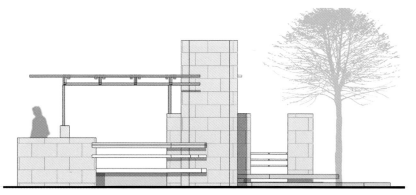

休息亭北立面

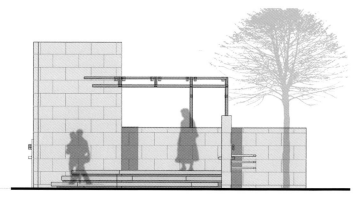

休息亭西立面

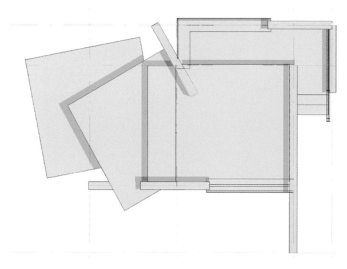

休息亭一层平面

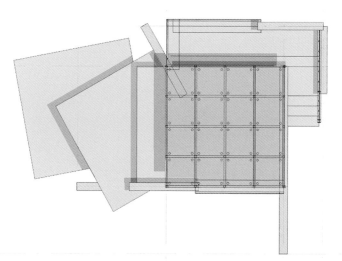

休息亭顶平面

休息亭实景照片

公共景观
PUBLIC LANDSCAPE

桐庐广场设计
PUBLIC SQUARE DESIGN, TONGLU

项目地点：浙江省桐庐县	Location: Tonglu, Zhejiang	项目成员
设计时间：2000 年	Design Period: 2000	景观设计：成玉宁、汪松陵
建成时间：2001 年	Completion Time: 2001	建筑设计：成玉宁
委托单位：桐庐县人民政府	Client: People's Government of Tonglu	结构设计：王伟成
用地面积：8 公顷	Area: 8 ha	水电设计：刘 俊、龚增谷

项目概况

桐庐广场位于富春江南岸，是桐庐新城跨江发展的核心。基地东西向长 800m、南北向宽 100 m，是由政府、办公建筑围合而成的城市中心绿地。桐庐历史悠久、人文荟萃，为景观设计提供了重要的文脉资源。

设计策略

城市道路将基地分隔为三部分，在遵循整体性原则的前提下，对三个片区采取"和而不同"的设计策略。

设计重点打造了东西主轴线，以联系三大片区；在保证主轴线完整性的基础上，强化南北向的沟通与变化，营造出收放开合、相互渗透的景观空间。轴线中段的空间适当向两侧山体延伸，形成景观次轴，并以树阵划分空间，在增加通透性的同时，形成了景观的空间序列与节奏。

设计着重协调广场与周边城市环境的对话关系。东部片区的景观构成与政府办公建筑相呼应；中部片区采用树阵与适当的隔离增加景观层次；西部片区以"桐庐之门"为主体，集中展现了桐庐的人文典故与历史风貌。

Project Profile
Located on the southern bank of Fuchun River, Tonglu Public Square is the core of the new city's cross-river development. This site's east-west length is 800m, and its south-north width is 100m. The square is the city's central greenbelt consisting of government and business buildings. Tonglu's long history and countless talents are the important context resources for the landscape design.

Design Strategy
The city roads divide the site into three sections. With the principle of integrity as the precondition, different strategies are adopted in the three sections, which seek harmony but also maintain diversity.

The design uses the construction of an east-west axis to connect the three sections. With the integrity of the main axis guaranteed, the connection is strengthened and some changes are added in the south-north section, which creates a landscape space with openness, connotation and transparency. The central space of the axis extends to the mountains on the both sides at a some degree, which forms a secondary axis of the landscape. Tree arrays are used to divide spaces, which not only adds the transparency, but also forms the spatial order and rhythm of the landscape.

The design focuses on coordinating the dialog relationship between the square and its surroundings. The landscape formation in the eastern section shows harmony with the government office building. Tree arrays are used in the central section to appropriately separate the east-west axis and add landscape levels. "Gate of Tonglu", the main part of the western section, provides a concentrated demonstration of Tonglu's cultural allusion and historical style.

桐庐广场中部鸟瞰实景

| 228 | PUBLIC LANDSCAPE   公共景观

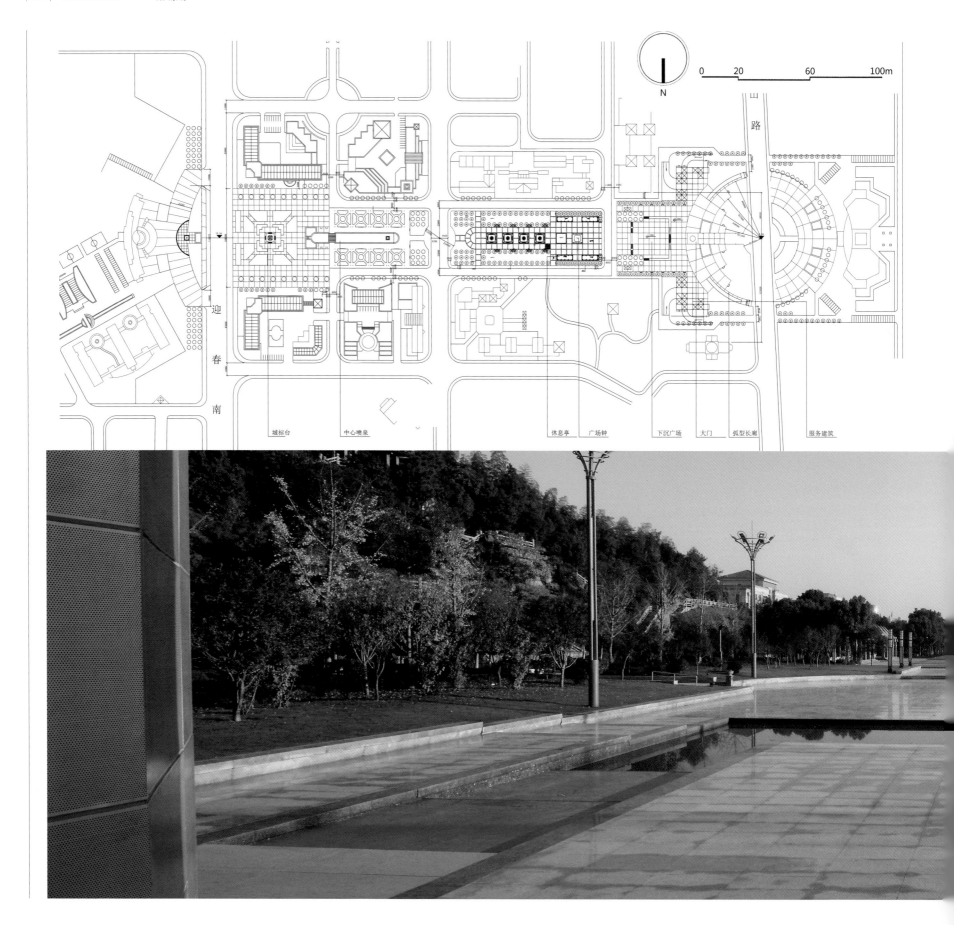

城标台　中心喷泉　　休息亭　广场钟　下沉广场　大门　弧型长廊　服务建筑

1. 桐庐广场总平面图
2. 广场轴线实景
3. 广场西段实景

桐庐之"门"实景照片

广场喷泉夜景

1. 桐庐之"门"及入口雕塑
2. 桐庐之"门"透视
3. 桐庐之"门"内部

公共景观　PUBLIC LANDSCAPE　| 233 |

1. 广场中段实景
2. 北部荷塘实景

广场银杏林实景

宿迁市运河金陵国际会议中心景观设计
JINLING GRAND CANAL INTERNATIONAL CONFERENCE CENTER AND HOTEL LANDSCAPE DESIGN, SUQIAN

项目地点：江苏省宿迁市	Location: Suqian, Jiangsu	项目成员
设计时间：2011 年	Design Period: 2011	景观设计：成玉宁、王卿卿、李志刚、赵曦志
建成时间：2013 年	Completion Time: 2013	水电设计：王晓晨、李滨海
委托单位：江苏运河文化城有限公司	Client: The Jiangsu Canal Cultural City Limited	
用地面积：25.9855 公顷	Area: 25.9855 ha	

项目概况

运河金陵国际会议中心位于宿迁市京杭运河以北、骆马湖南部。会议中心集国宾接待、政务活动、高端商务、休闲度假和国际会议等综合服务功能于一体。场地呈外高内低的向心之势，设计因地就势，梳理竖向关系，营造了具有"山水"意趣的酒店景观。

设计策略

设计妥善处理了建筑与场地之间的竖向矛盾，在就地平衡土方的基础上营造山水地貌，并利用种植强化竖向变化，创造出"丘陵平湖"的空间效果，不仅丰富了会议中心景观环境，还满足了行洪的要求。设计利用植物材料柔化建筑与外部环境的边界，在道路两侧采用群落式配植，形成了林木深深的景观效应。

湖面是园区景观的中心。西片区总统楼四周环绕以带状水系，祥和静谧，且保证了私密性；北片区结合水景点缀假山栈道，主楼前布置草坪。北入口片区以台地、莲花无界水池结合"流觞台"，唤起人们对古运河文化的记忆；东部片区岗埠逶迤，仿佛山间；南部片区进门见湖，隔湖相望主体建筑端庄而不失清秀……不一样的运河金陵国际会议中心应运而生。

Project Profile

Jinling Grand Canal International Conference Center lies to the north of the Beijing-Hangzhou Canal and to the south the Luoma Lake. The conference center can fulfill a complete list of service functions for state guest reception, state affairs, high-end business, recreation and international conferences. Since the external section is higher than the internal, this site is at a centripetal state. The design utilizes this feature to sort out the vertical space, thus creating a hotel landscape with the artistic interest of "mountain and water".

Design Strategy

This design takes appropriate measures to resolve the vertical contradiction between architectures and the site. Landforms of mountains and rivers are created on the basis of balancing the earthwork on the spot. Plantation is used to intensify the vertical changes so as to create the spatial effect of slack water lying in hills. The design not only enriches the landscape environment of the conference center, but also meets the requirements of flood drainage. Plant materials are used to soften the border between architectures and the external environment. Plant communities are planted on both sides of the road, which creates a landscape image of deep and serene forests.

The lake surface is the landscape center of the park. In the western section, the president building is surrounded by belt-shaped water system, which not only creates tranquil atmosphere but also can keep the privacy. In the northern section, rockeries and plank roads are dotted with the water landscape, and meadows are arranged in front of the main building. In the northern entrance section, platforms and the Lotus Boundless Pond along with the "Liushang Terrace" recall people's memories of the old canal culture. In the eastern section, the winding docks looks like zigzag mountain paths. In the southern section, lake can be seen as soon as entering the gate, and the main buildings on the opposite side are both dignified and delicate… Distinct and unique, the Jinling Grand Canal International Conference Center emerges as the times require.

公共景观　PUBLIC LANDSCAPE

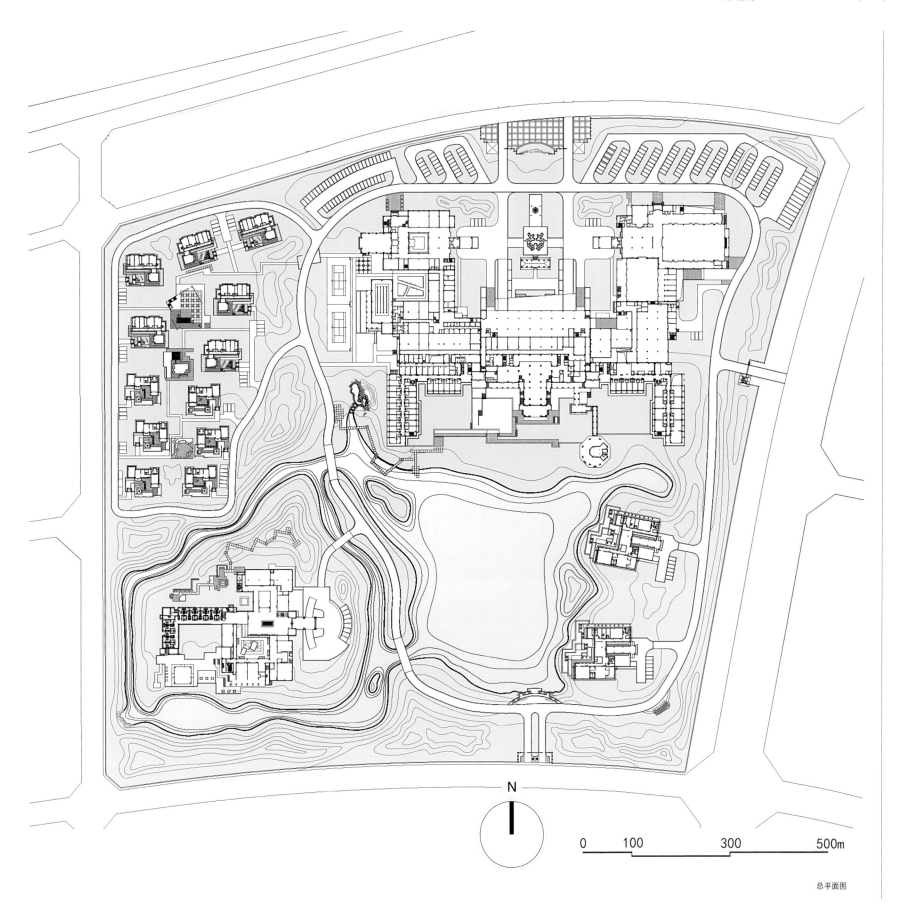

总平面图

1	
3	1. 酒店鸟瞰图
2	2. 北入口景墙实景
	3. 北入口无边界水池实景

北入口广场中轴景观

1. 假山跌水实景
2. 溪流栈桥实景

中央湖区景观

1. 云栖竹径实景
2. 回桥实景
3. 酒店庆典草坪实景

总统楼环抱水系实景

园桥实景

南京市金陵江滨酒店景观设计
JINLING RIVERSIDE HOTEL LANDSCAPE DESIGN, NANJING

项目地点：江苏省南京市	Location: Nanjing, Jiangsu	项目成员
设计时间：2013年	Design Period: 2013	景观设计：成玉宁、陈烨、赵曦志、刘常宁
建成时间：2014年	Completion Time: 2014	建筑设计：成玉宁、陈烨、黄元圣
委托单位：南京河西新城区开发建设指挥部	Client: Nanjing hexi new town development and construction headquarters	结构设计：马军
用地面积：60451.2平方米	Area: 60451.2m²	水电设计：王晓晨、李滨海

项目概况

金陵江滨酒店位于绿博园之中，毗邻南京奥体中心。设计梳理、整合了现状资源，重组景园空间以提升环境品质，营造出与酒店定位及风格相契合的景观氛围。

设计策略

协调庭园空间与建筑的对话关系，结合酒店建筑布局多个主题园区，拓展户外活动空间，利用空间的收放开合，丰富酒店的景观环境。设计有选择地保留长势良好的树木为基调植物种，采取"同类项合并"的方式，将原本混栽的树木归类以形成主题园，增植宿根和彩化地被植物，营造精致的植物景观。

设计重新组织园区交通，优化道路及广场铺装，结合人车分流，丰富了铺装类型。设计将园区与外部水系相沟通，改善了内湖水质，通过梳理，形成湖、溪、池、瀑等不同类型的水景。并且，对水岸进行了优化与调整，形成缓坡、浅滩，间或栽植菖蒲、鸢尾等水生植物，营造了自然优美的滨水岸线。

Project Profile

Seated in the Green Expo Garden, Jinling Riverside Hotel is adjacent to Nanjing Olympic Center. This project sorts out and integrates the current resources, and reorganizes landscape spaces to improve the environment quality, thus creating a landscape atmosphere that is in line with the hotel's orientation and style.

Design Strategy

The design coordinates the dialog relationship between the park and the architecture, and arranges several theme parks among the hotel buildings so as to expand its space for outdoor activities. It also utilizes the openness and connotation of the space to enrich the hotel's landscape environment. Trees which grow well are selected to be maintained as the keynote, and the companion planting trees are categorized by the method of "combining like items" for the purpose of constructing a theme park. Exquisite plant landscape is created by adding perennial plants and coloring cover plants.

This design reorganizes the traffic within the park. It optimizes roads and the square pavement, and enriches the pavement types through separating the road for pedestrians and vehicles respectively. The design connects the internal water with the external water system, thus improving the water quality of the internal lake. Through classification and integration, different types of water landscape are formed, such as lake, stream, pond and waterfall. The waterfront is optimized and adjusted, which bring the formation of gentle slopes and shoals. Water plants like calamus and iris are alternatively planted, creating a natural and elegant waterside shoreline.

公共景观 | PUBLIC LANDSCAPE | 253

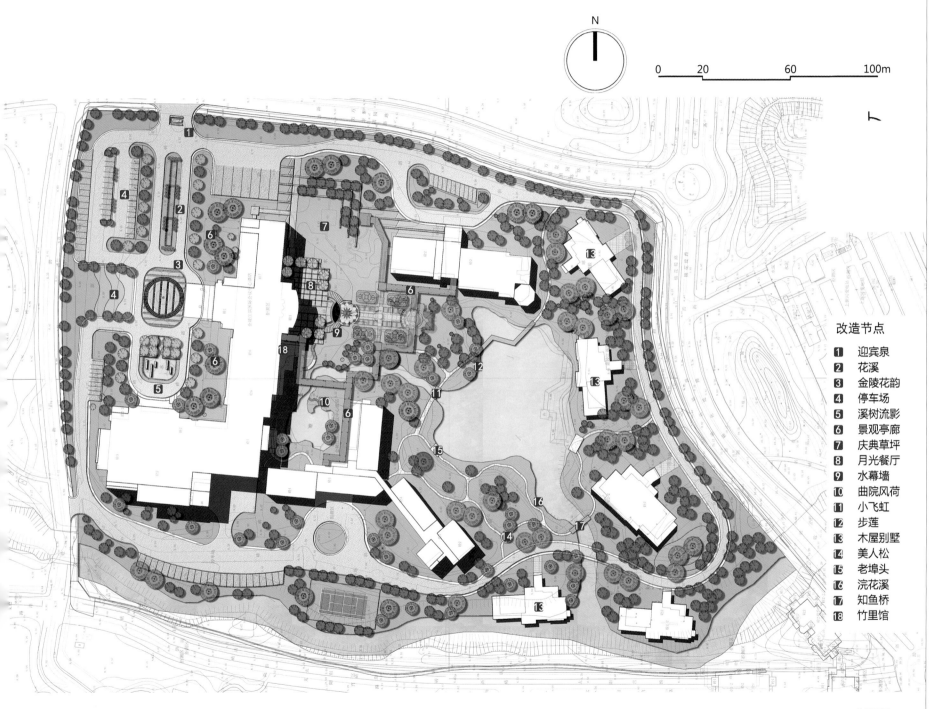

改造节点

1. 迎宾泉
2. 花溪
3. 金陵花韵
4. 停车场
5. 溪树流影
6. 景观亭廊
7. 庆典草坪
8. 月光餐厅
9. 水幕墙
10. 曲院风荷
11. 小飞虹
12. 步莲
13. 木屋别墅
14. 美人松
15. 老埠头
16. 浣花溪
17. 知鱼桥
18. 竹里馆

总平面图

酒店主楼入口实景

1. 花溪实景
2. 迎宾泉实景
3. 金陵花韵实景

酒店大堂对景

酒店庭院

庆典草坪实景

1. 中庭跌水实景
2. 庆典草坪挡墙细部
3. 曲院风荷实景

1. 园路实景
2. 游步道实景

1. 草坪景观
2. 知鱼桥
3. 步莲亭

中心湖区景观

| 270 | PUBLIC LANDSCAPE   公共景观

1. 园内步道
2. 疏林小径
3. 疏林草地
4. 曲径通幽
5. 步莲亭曲桥

1. 林间石径
2. 木栈桥
3. 濠濮间想

南京市新街口正洪广场景观环境改造设计
XINJIEKOU ZHENGHONG SQUARE LANDSCAPE REDESIGN, NANJING

项目地点：江苏省南京市	Location: Nanjing, Jiangsu	项目成员
设计时间：2014年	Design Period: 2014	景观设计：成玉宁、李 哲、谭 明、鲍洁敏、
建成时间：2014年	Completion Time: 2014	单梦婷、濮岳川、赵玉龙、刘 悦
委托单位：南京新街口商业步行街管理办公室	Client: Nanjing xinjiekou commercial pedestrian street management office	水电设计：王晓晨、李滨海
用地面积：18322平方米	Area: 18322m^2	

项目概况

新街口商业步行街位于南京市核心地段，不仅是购物消费的汇金之地，也是市民休闲娱乐的文化场所，更是展示城市魅力的重要窗口。新街口商业步行街以正洪广场为中心，在东、南、西、北四个方向延伸出四条主要商业街道，均与城市干道相连，交通便捷、人流量大。设计更新了街区的景观环境，让南京的"城市客厅"展现出充满生机的新景象。

设计策略

设计提出"一心四轴"的空间格局："一心"为中心商业广场，设计充分考虑其作为集散广场的要求，保留了空间的整体性与开敞性，以铺装的肌理呼应周边的建筑。四条步行街道作为"四轴"，设计在整合公共空间的基础上，增设造型简洁、易维护的白麻石花坛坐凳。改造后的中心绿化带两侧各有不少于6米宽的通道，既满足了功能需求，又从视觉上协调了步行街和周边建筑的尺度，缓解了高层建筑带来的压迫感。

设计引入了"四方神兽"与"五色土"等传统文化元素，以赤、黑、青、白四色区分不同区域，增加"四轴"视觉可识别性；选取入口醒目位置嵌入"四方神兽"地雕，作为出入口的标识；并于四条步行街中，分别种植红果冬青、杜英、桂花和香樟，进一步强化景观特征。新增的雾森和直饮水机等公共设施，使景观环境更具人性化。景观亮化系统在提供夜间照明的同时，形成了独特的夜景效果。

Project Profile

Located in the core section of Nanjing, Xinjiekou Zhenghong Square is not only a shopping center but also a culture site that can meet citizens' recreation needs. Furthermore, it is an important window from which the city's charm is displayed. With Zhenghong Square as the center, four business streets are the outreaches of Xinjiekou Zhenghong Pedestrian Streets to four directions, namely east, south, west, and north. These four streets are connected to the city's arteries, which brings convenient transportation and large pedestrian volume. This design updates the landscape environment of the streets, making Nanjing's "living room" take on a new vigorous image.

Design Strategy

The spatial pattern put forward by the design is called "one core, four axes". Here, "one core" refers to the central business square. Since the design takes full consideration of the requirements of being an evacuation square, the space's integrity and openness are maintained, with pavement texture assimilated with the surrounding architectures. "Four axes" are the four pedestrian streets, the design of which adds brief-styled and easily-maintained white granite parterre stools based on the integration of public spaces. After the reconstruction, two passageways over 6 meters wide are added on both sides of the central greenbelt, which not only meet functional demands, but also visually coordinate the scale of the pedestrian streets and the surrounding buildings, thus remitting the constriction caused by high-rise buildings.

The design introduces traditional cultural elements like "Mythical Creatures in Four Directions" and "Five-color Flower". Red, black, blue and white are four colors that represents four divided sections, which strengthen the identity of the "four axes". The floor sculptures of "Mythical Creatures in Four Directions" are inserted into evident positions near the entrance so as to mark the entrance. In the four streets, holly, Elaeocarpus decipiens, osmanthus trees, and camphor trees are planted to strengthen landscape features. Public facilities like spray and direct drinking fountain are added to make landscape environment more humane. The landscape lighting system not only provides night illumination, but also creates a unique night scape.

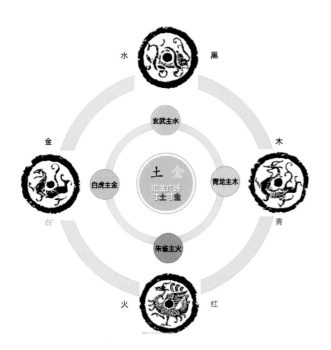

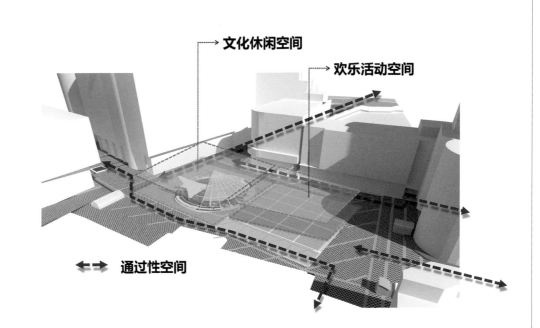

主题色彩的铺装

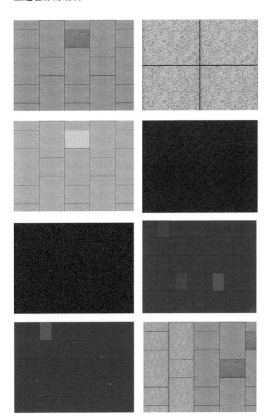

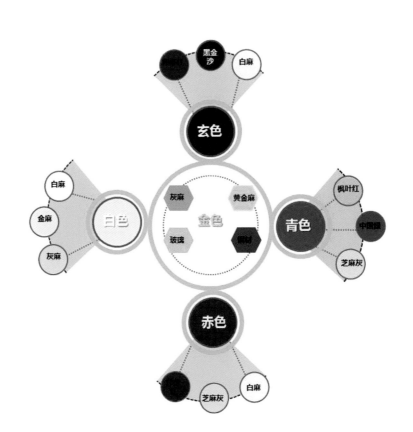

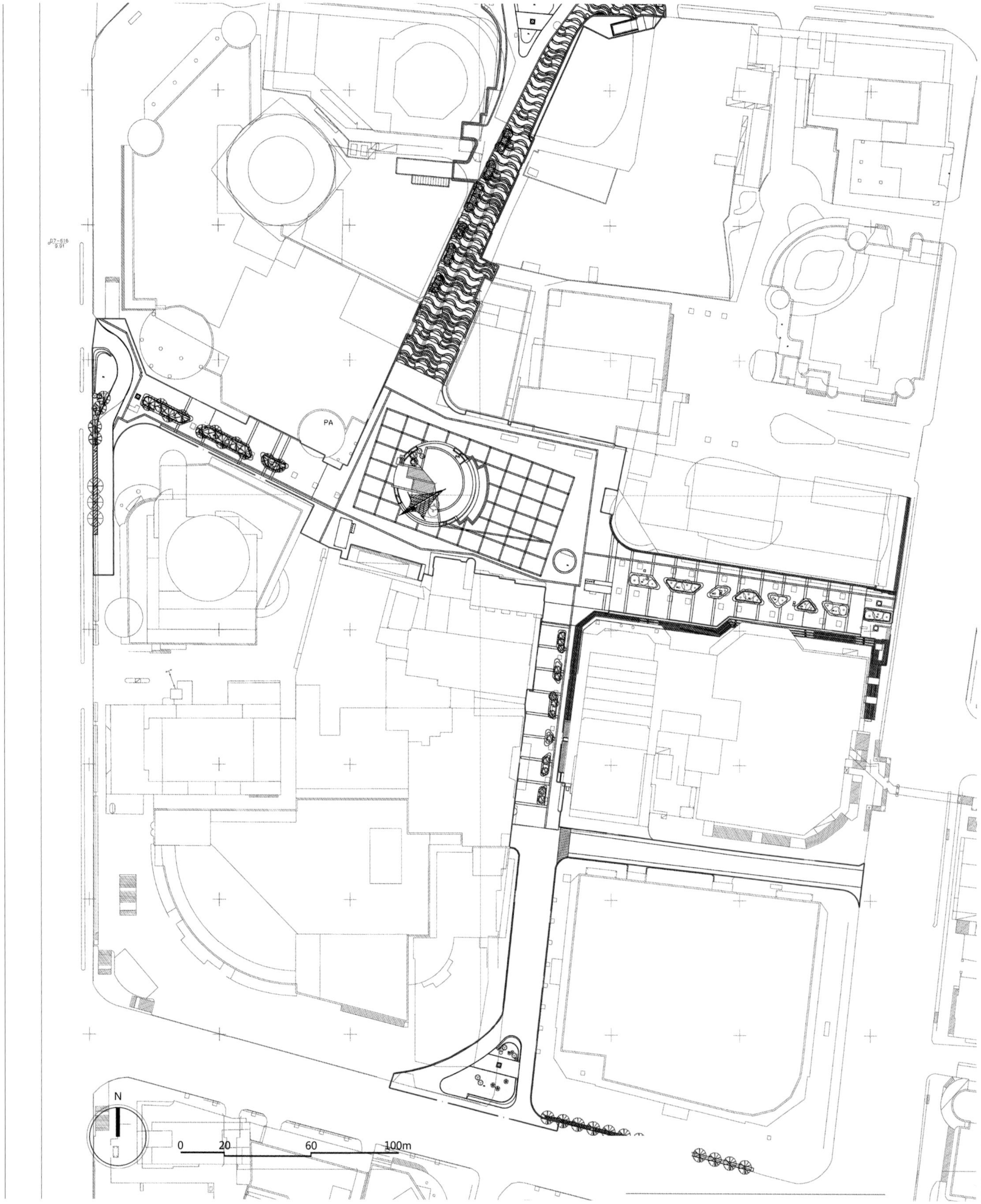

| 276 | PUBLIC LANDSCAPE 公共景观

0 20 60 100m

1. 新街口总平面图
2. 正洪广场平面图
3. 正洪广场实景

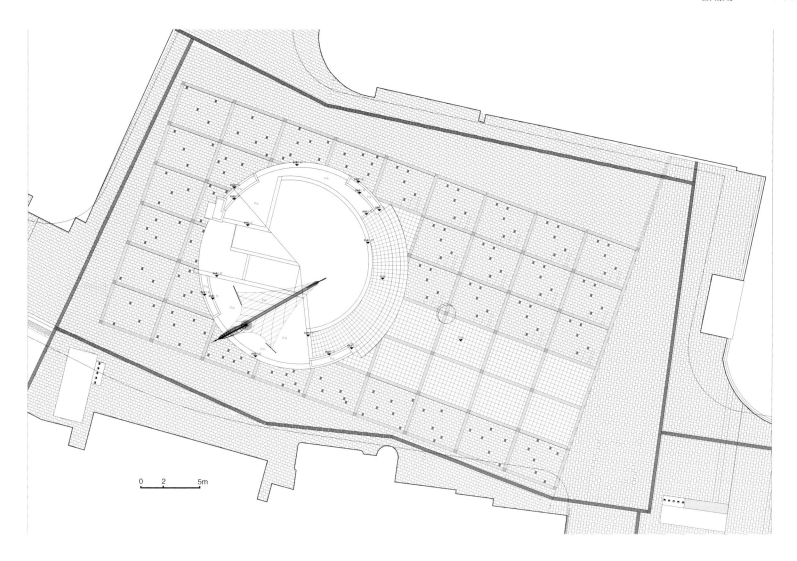

正洪广场雾森

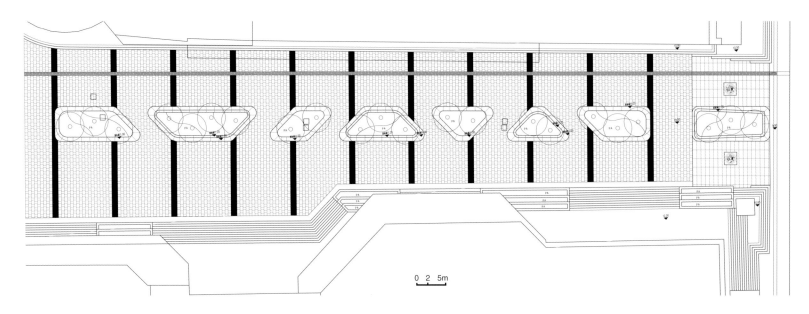

1. 东入口广场平面图
2. 东入口广场雾森实景
3. 东入口广场景观坐凳实景

| 282 | PUBLIC LANDSCAPE 公共景观

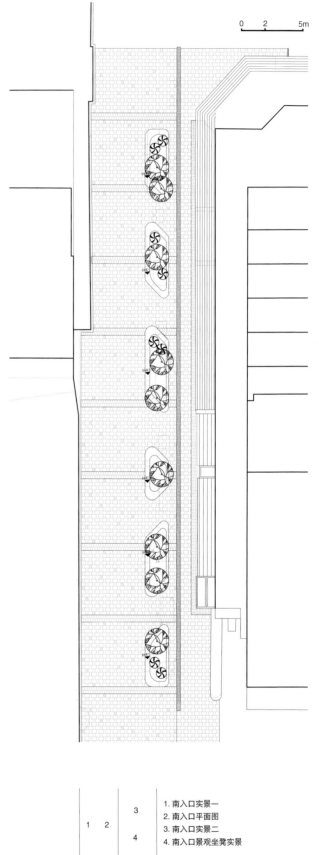

| | | 3 | 1. 南入口实景一
| 1 | 2 | | 2. 南入口平面图
| | | 4 | 3. 南入口实景二
| | | | 4. 南入口景观坐凳实景

| 284 | PUBLIC LANDSCAPE 公共景观

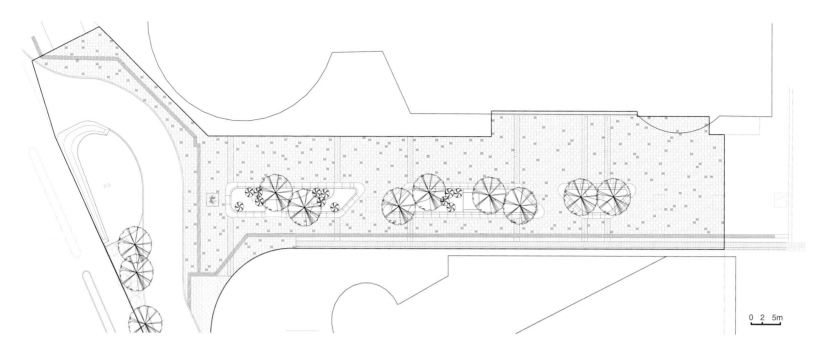

1. 西入口广场平面图
2. 西入口景观实景一
3. 西入口景观实景二

1. 西入口步行街实景
2. 西入口植坛实景

公共景观 | PUBLIC LANDSCAPE

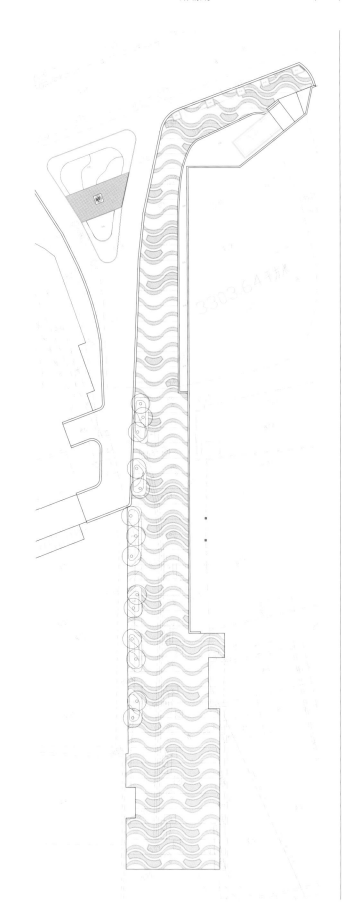

1. 北入口实景
2. 北入口平面图

住区景观
RESIDENTIAL LANDSCAPE

南京市新世界花园景观设计
NEW WORLD GARDEN LANDSCAPE DESIGN, NANJING

项 目 地 点：江苏省南京市	Location: Nanjing, Jiangsu	项 目 成 员
设 计 时 间：2005 年	Design Period: 2005	景观设计：成玉宁、杨芳龙
建 成 时 间：2006 年	Completion Time: 2006	水电设计：刘 俊、刘向上
委 托 单 位：南京新丽都房地产开发有限公司	Client: New Lidu real estate development Co., Ltd. Nanjing	
用 地 面 积：58273 平方米	Area: 58273 m²	

项目概况

新世界花园位于南京紫金山麓、玄武湖畔。设计以"都市山居"为理念，充分结合山地环境条件，采用地带性植物种，彰显外部环境特征，营造符合现代都市生活需求的生态型住区环境。

设计策略

设计因地制宜、延山点景，融住区于紫金山大环境之中，以凸显山居文化、山林意趣。住区景观空间由五个区域构成：A 区毗邻宁栖公路，为弱化交通影响而大面积栽植竹林，形成绿色屏障，统一沿街景观；B 区作为游憩场所，与 A 区在视觉上相互渗透，形成视觉通廊，可眺望紫金山景；中心绿化带充分利用了条状台地空间，成为健身活动区；中心花园植林构亭，为居民提供富有文化韵味的休闲场所；别墅区的种植以大乔木为主，配以疏林草地……。

在充分研究现状，重新组织住区交通，外围设置环路，最大程度地分流人车、增加景观绿地的面积，丰富景观环境。建成后的住区内部绿树参天、浓荫如盖，为人们提供了多样化的户外休憩活动空间。

Project Profile
New World Garden is located on the foot of Purple Mountain and beside Xuanwu Lake in Nanjing. Inspired by the idea of "mountain life in the urban areas", the design fully integrates the environment conditions of mountain, and adopts zonal plant species, so as to highlight the external environment features and create an ecological living environment which meets the needs of modern urban life.

Design Strategy
The design adjusts to local conditions, extends the mountain and intersperses scenes. Residential quarters are integrated into the overall environment of Purple Mountain, which highlights the culture of living in the mountain and the artistic interest of mountain and forests. The landscape space of residential quarters consists of five sections. Section A is adjacent to Ningqi Road. To reduce the effect caused by traffics, bamboo forests are planted in a large area, which forms a green defense and unifies the landscape along the streets. Section B is used as a rest and recreation site. Section B and Section A are visually interpenetrative, which forms a visual corridor for overlooking the scenery of Purple Mountain. The third section is the central greenbelt which takes full advantages of the belt-shaped platform space to function as a body-building zone. The forth section is the central garden, where trees are planted and pavilions are built, offering citizens a recreation site with cultural appeals. The last section is the villa zone. The major plants of this section are magaphanerophytes, with open forest and grassland dotted in some areas…

With the current conditions fully studied, traffics in the residential quarters are redesigned; contours are arranged on the periphery; roads for pedestrian and vehicles are separated to the largest extent, the area of green land is increased, and the landscape environment is enriched. After the construction, green trees towering in the residential quarters and bringing dense shades, people are provided with diverse spaces for outdoor activities.

新世界花园入口效果图

延紫金山脉
聚金陵之气
营山居氛围
显仁者风范

1. 总平面图
2. 鸟瞰图

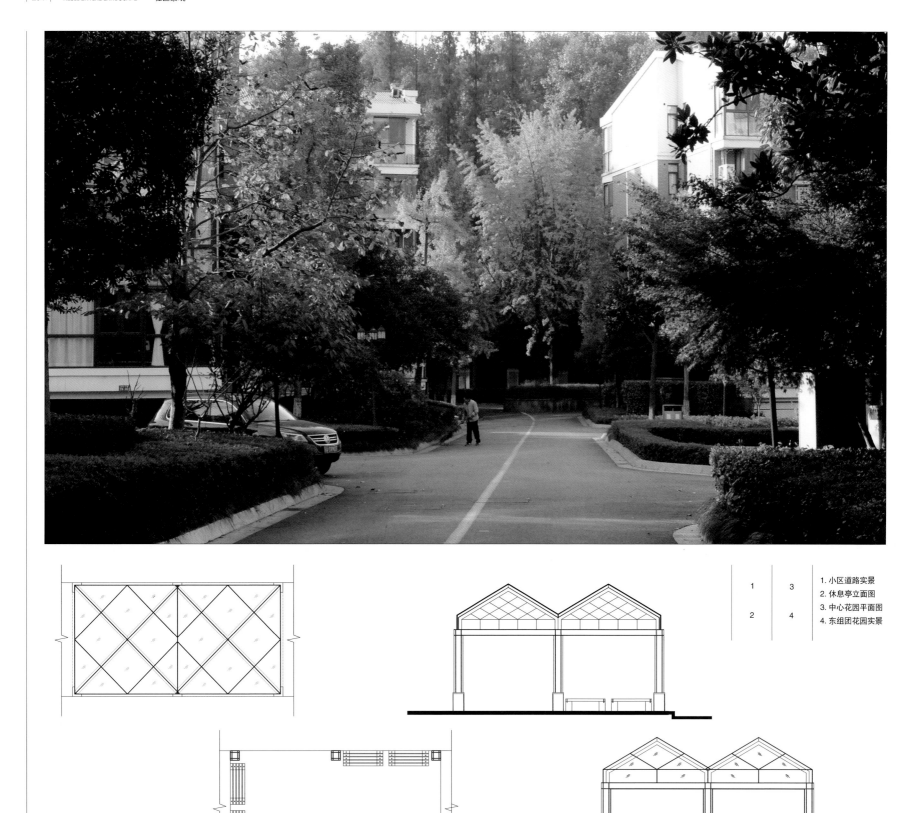

1. 小区道路实景
2. 休息亭立面图
3. 中心花园平面图
4. 东组团花园实景

中心花园入口实景

1. 儿童天地实景
2. 小区园路实景
3. 东花园鸟瞰

1. 宅间实景一
2. 宅间实景二
3. 中心花园实景

1. 宅间实景
2. 园路实景
3. 云溪竹径

武夷山市华彩家园景观设计
LANDSCAPE DESIGN OF HUACAI JIAYUAN IN WUYISHAN CITY

项 目 地 点： 福建省武夷山市	Location: Wuyishan, Fujian	项 目 成 员
设 计 时 间： 2008 年	Design Period: 2008	景观设计： 成玉宁、张 祎、许立南、张亚伟
建 成 时 间： 2009 年	Completion Time: 2009	水电设计： 王晓晨、李滨海
委 托 单 位： 武夷山市霈耕房地产开发有限公司	Client: Peigeng real estate development Co., Ltd. Wuyishan	
用 地 面 积： 30071 平方米	Area: 30071m²	

项目概况

华彩家园位于福建省武夷山度假区。设计以武夷山为背景，小中见大地再现了武夷山水意境，营造出人性化、地域特色鲜明的住区环境。

设计策略

该项目基地高差大、用地紧凑，秉承"久在城市里，复得返自然"的理念，设计以"武夷九区溪"为景观设计意境，因高就低，于住区中形成东西与南北两条水轴，并结合宅间绿地等留出多条视线通廊，以远借武夷山景。远山近水尽显山水意趣。

景观结构上分成"二轴、三片"。"二轴"中的主轴自东入口回环至西入口，依次布置溪流、池塘、跌水、涌泉、石矶、平台等节点；南北向的景观辅轴充分利用了住宅山墙间的开放空间，曲水萦绕宅间，点缀布置景观小品及观景平台，形成中轴景观序列。另外，依据东、西两坡及丘顶将景观分作三个片区，中部为水景，东、西两坡采用台地结合盘山园路，景观空间流畅，层次丰富。此外，设计巧妙利用地形高差、在满足私密性的同时将公共区域景观与住户庭园、停车场空间等相融合，有效地扩大了住区外部景观空间。设计利用本山的红砂岩作为挡墙材料，富有地域特色。

Project Profile

Huacai Jiayuan is located within the Wuyi Mountain Resort in Fujian province. The design, utilizing Wuyi Mountain as its background, displays the artistic landscape of Wuyi Mountain, which creates a people-oriented residential environment with distinctive regional characteristics.

Design Strategy

The land of the project site is compact and has large elevation difference. This design, sticking to the idea of "returning to the nature is necessary after a long stay in the city", utilizes "Jiuqu Stream" as the artistic conception of the landscape design. Adjusting to the elevation difference, the design forms two axes, namely the west-east axis and the south-north axis. Through integrating the green land between houses, several visual corridors are created, where a distant view of Wuyi Mountain can be appreciated. With close water and distant mountain, the designed landscape is full of interest and charm.

The landscape structure consists of "two axes and three sections". The main axis of "two axes" starts from the east entrance and ends with the west entrance, along which landscape nodes like stream, pond, head fall, spring and terrace are sequentially arranged. The south-north auxiliary axis gives a full play of the open space between houses and gable walls, which creates an axial landscape array, where water winds along the houses, and accessorial buildings and observation platforms are set as embellishment. In addition, according to the west and east braes as well as the knap, the landscape is divided into three sections. Waterscape is arranged in the central section, while bench terraces are integrated into the hilly road. In this way, the landscape enjoys a smooth space and rich gradation. Besides, the design tactfully uses the elevation difference, and meets demand of privacy while effectively expands the landscape space outside the residential section through integrating landscape of various sections, such as the public section, the residential section and the parking space. Red sandstone from Wuyi Mountain is used as the parapet material, which also shows distinctive regional features.

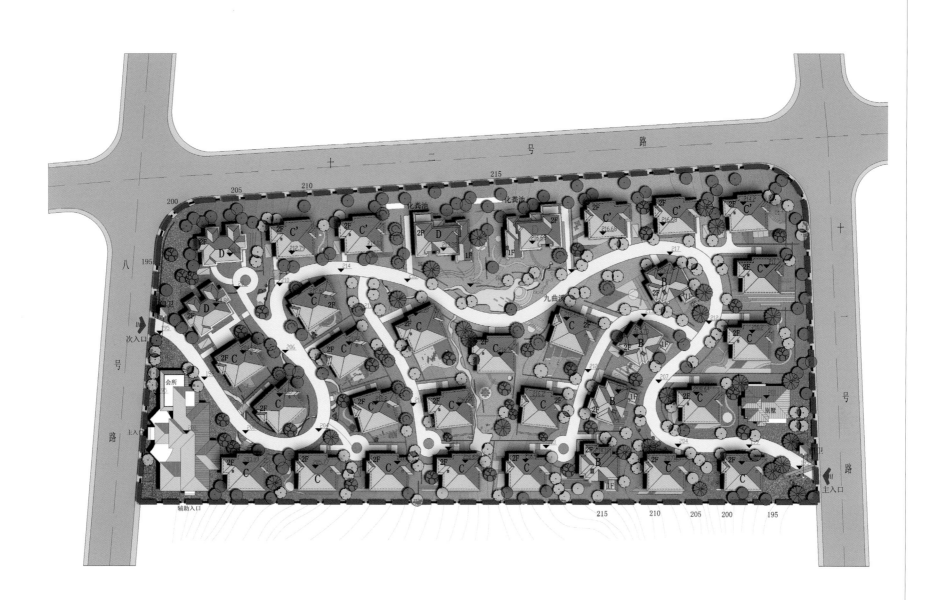

总平面图

小九曲溪水景

1. 九曲溪水景透视一
2. 九曲溪水景透视二
3. 园路透视

住区景观　RESIDENTIAL LANDSCAPE

1. 池畔别墅实景
2. 园路实景
3. 九曲溪溪流实景
4. 挡墙植坛实景

1. 园路实景
2. 红砂岩挡墙实景
3. 入户道实景

九曲溪滨水平台实景

景观建筑
LANDSCAPE&ARCHITECTURE

南京市老山森林公园景观及建筑设计
ARCHITECTURE AND LANDSCAPE DESIGN OF LAOSHAN FOREST PARK, NANJING

项目地点：江苏省南京市	Location: Nanjing, Jiangsu	项目成员
设计时间：1998年	Design Period: 1998	景观设计：成玉宁、陈烨、王巍巍、苏雅茜
建成时间：2000年	Completion Time: 2000	建筑设计：成玉宁、陈烨、李竹青、张梦薇、曾明理
委托单位：南京老山森林公园管委会	Client: Nanjing laoshan forest park management committee	结构设计：马军、赵才其
用地面积：8000公顷	Area: 80 ha	水电设计：王晓晨、李滨海

项目概况

老山国家森林公园位于南京长江北岸，是江北重要的生态林地。设计以生态保护为原则，全面整合园区资源，提升景区旅游服务设施品质。

设计策略

设计在维护生态特色的基础上，进一步提升了森林公园的景观效果，注重地域文化的表达，在延续地域文脉的同时满足了人们休闲娱乐的需求。

七佛寺景区黄山岭入口的改造与塌方段整治工程相结合，统筹设计，利用现状地形突出了主入口的可识别特征。入口大门采用钢结构，覆以芦苇屋面，整体形象如起伏的山峦，质朴而清新。入口西部设有餐饮中心，采取院落式布局，因山就势，融入山林。依托公园内的人文资源遗存，结合现有的张孝祥墓和塑像打造状元文化，设置状元广场、书山有路（登山道）、三元及第三道牌坊、许愿台、老鹰山状元台等景点，形成彰显状元文化的景观游线，吸引莘莘学子前来体验浓厚的尚学传统。

公园入口景观标识

Project Profile
Laoshan National Forest Park is an important ecological forest located on the north bank of Yangtze River. With ecological protection as the principle, this design integrates the overall resources and improves the quality of tourism service facilities in the park.

Design Strategy
On the basis of maintaining the ecological features, the design further improves the landscape effect of the forest park. With emphasis put on the expression of regional cultures, the design sustains the regional context and meets people's recreation demands at the same time.

The Huangshanling entrance of Qifo Temple is reconstructed, which combines the renovation project in the collapsing sections. The two projects are planned as a whole, utilizing the current landform to highlight the identity of the main entrance. The entrance gate uses steel structure and is covered by bamboo reeds, the image of which is like undulating hills and brings plain and pure feelings. The eastern part of the entrance lies the catering center. Utilizing courtyard patterns, it is built along the mountain so as to be naturally assimilated into the forests. The culture of Zhungyuan (title conferred on the one who came first in the highest imperial examination) is established. With the existing Zhang Xiaoxiang Tomb and his sculptures, more scenes that rely on the park's cultural resources are added, such as Zhungyuan Square, Shushan Youlu (mountain trail), three memorial archways for Sanyuanjidi (literally scholar who came first in three imperial examinations), Wishing Terrace, and Zhuangyuan Terrace. The culture of Zhuangyuan is highlighted in these landscape arrays, which attracts many students to come and experience the study tradition.

1. 入口片区总平面图
2. 餐饮中心

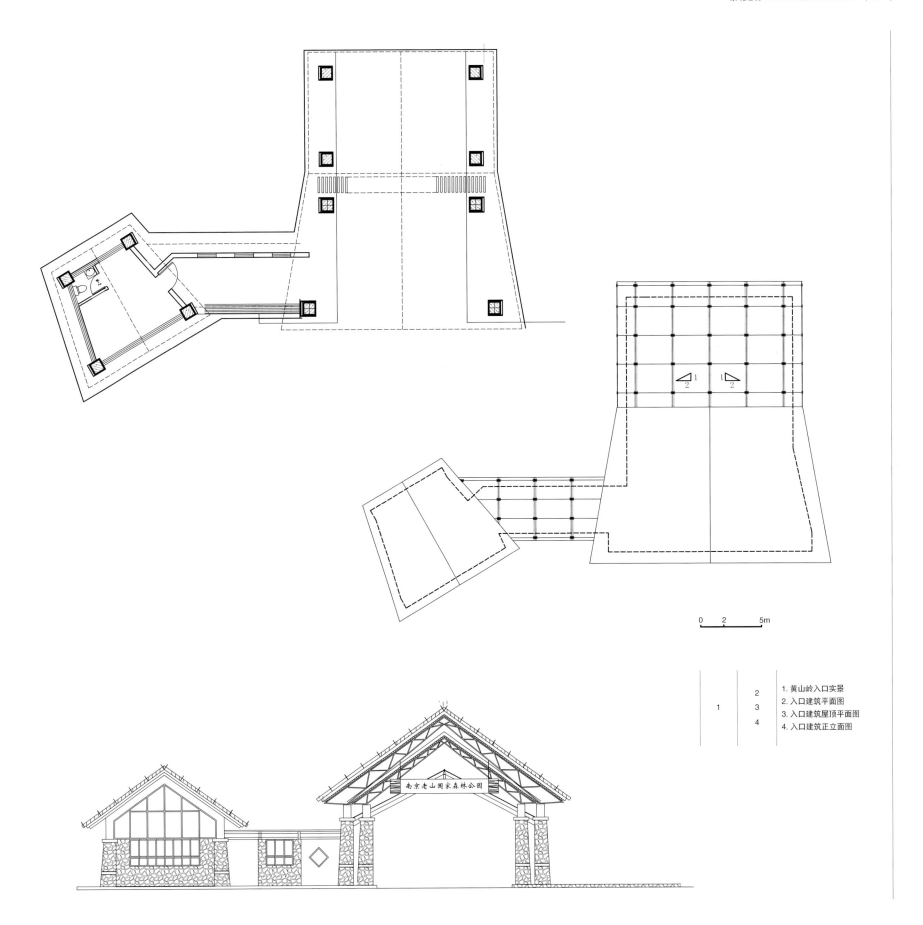

1. 黄山岭入口实景
2. 入口建筑平面图
3. 入口建筑屋顶平面图
4. 入口建筑正立面图

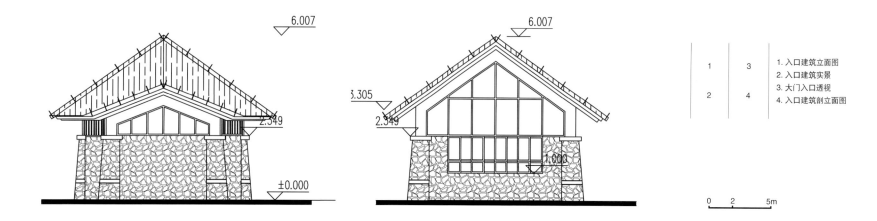

1. 入口建筑立面图
2. 入口建筑实景
3. 大门入口透视
4. 入口建筑剖立面图

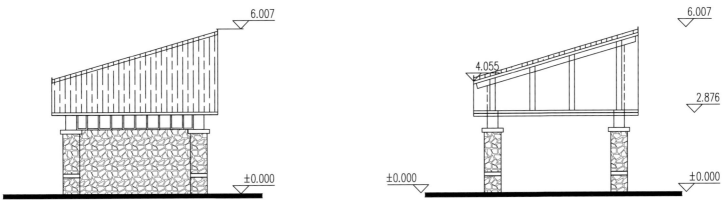

1. 黄山岭入口水景
2. 入口建筑立面、剖面图

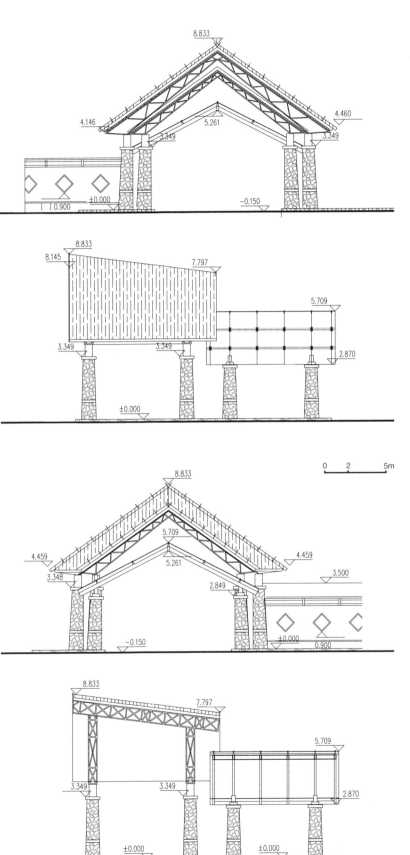

1. 景区入口透视
2. 入口广场实景一
3. 入口广场实景二
4. 清风长廊景观建筑实景

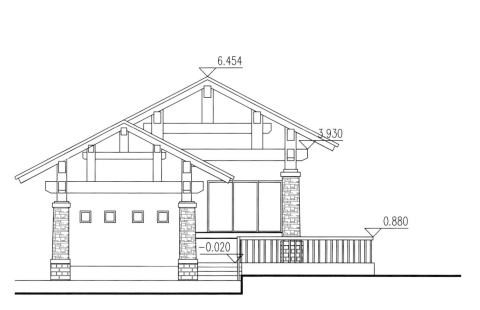
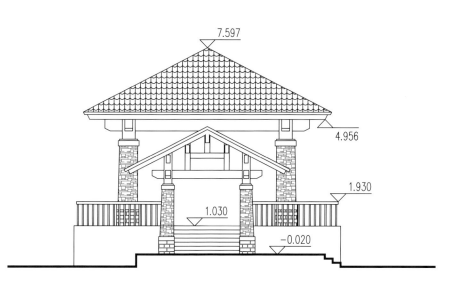

1. 清风长廊实景
2. 清风长廊剖面图
3. 清风长廊立面图

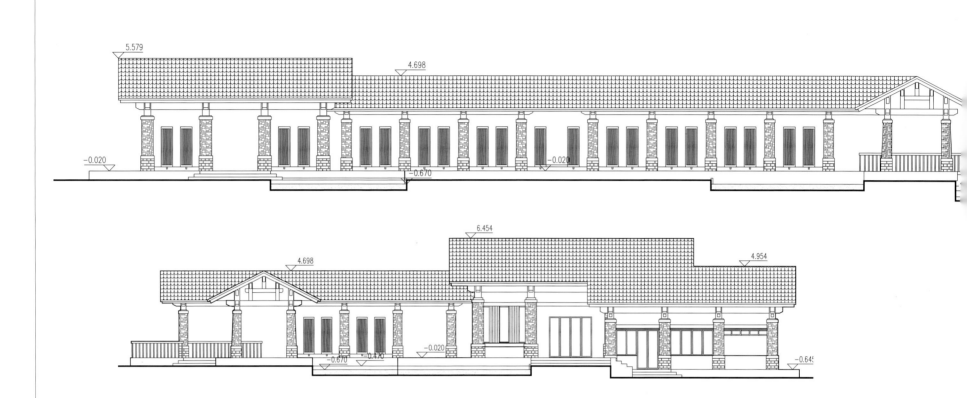
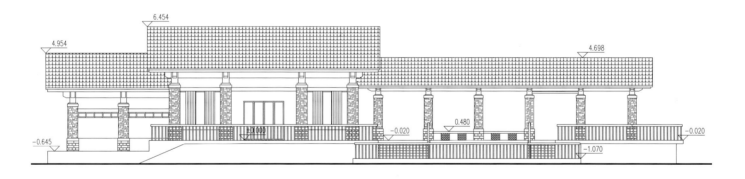
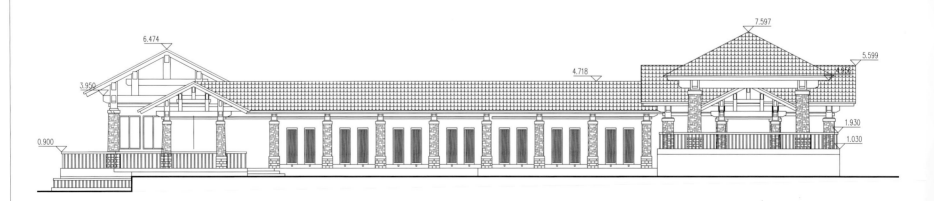

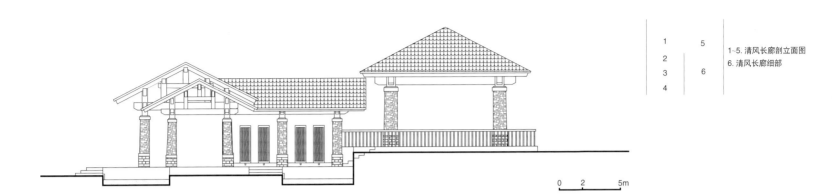

1-5. 清风长廊剖立面图
6. 清风长廊细部

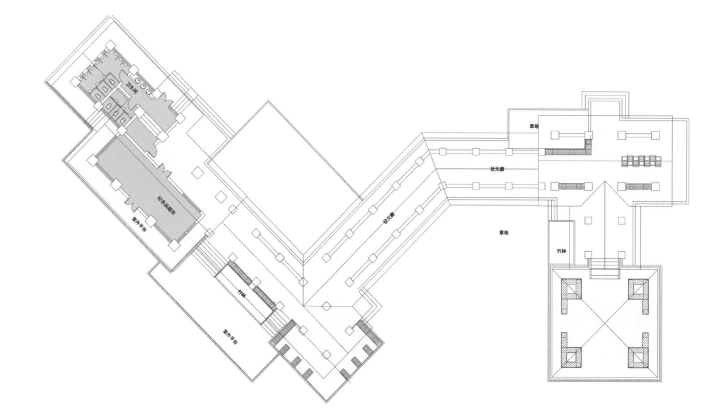

1		3	1. 清风长廊平面图
			2. 清风长廊局部
			3. 亭廊透视
2	4	5	4. 状元广场实景
			5. 栈道平台实景

1. 山间亭廊
2. 状元及第坊

南京市地铁一号线地面站站厅及景观设计
METRO LINE 1 GROUND STATION ARCHITECTURE AND LANDSCAPE DESIGN, NANJING

项目地点：江苏省南京市	Location: Nanjing, Jiangsu	项目成员
设计时间：2003 年	Design Period: 2003	景观设计：成玉宁
建成时间：2005 年	Completion Time: 2005	建筑设计：成玉宁、张楷
委托单位：南京市地铁建设指挥部	Client: Nanjing metro construction headquarters	结构设计：浙江精工、徐州新天地钢构
线路全长：16.99 千米	Area: 16.99 km	

项目概况

南京市地铁一号线一期工程构建了连通主城区中轴线的快速轨道交通。线路全长 16.99 千米，其中地下线 10.62 千米，地上线 6.37 千米。地面站自南向北分别为奥体中心站、小行站、安德门站、中华门站、红山动物园站及迈皋桥站，均处主城之内，人流量较大，空间环境复杂。设计对各个站点进行了充分的研究，在满足交通功能的同时，结合所处环境塑造了各具特色的站台建筑及景观。

设计策略

坚持站点环境与城市空间相协调的理念，变"叠加"为"楔入"，突出场所整体共生的原则，形成了独具特色的城市交通景观。

1）延续文脉的设计策略

设计突破了一般交通性广场所具有的"重交通轻文化"的设计模式，充分利用人文景观资源，凸显古城南京的人文内涵，营造出富有时代气息的地段性标志景观，体现城市新风貌。

2）人性化的场所空间营造

设计突出"以人为本"，最大限度地构建人性化的站前广场空间，在满足交通需要的基础上兼顾周边居民生活、休闲和交往的需求。

3）可持续的场所景观设计

秉承集约化的设计方法，力图构建一个低造价、高效益、可持续的地铁景观环境。设施及小品的设计遵循低维护原则；大量使用地带性植物，以落叶乔木为主，不仅满足夏季遮荫、冬季采光的需要，而且保证了空间的通透性。

Project Profile

Nanjing's metro line 1 construction project has realized the high-speed rail transit that links the axes of the main urban areas. The total main line track is 16.99km, of which 10.62km runs underground and 6.37km on or above the ground. From south to the north, the ground stations include Olympic Center, Xiaohang, Andemen, Zhonghuamen, Hongshan Zoo and Maigaoqiao. All the stations are located within the main urban areas, so the pedestrian volume is big and the spatial environment is complex. Having given all-round studies on each station, this design fulfills the site's transportation functions, and also creates distinctive station architectures based on the surrounding environment.

Design Strategy

Sticking to the idea of coordinating the station environment and the city space, the design utilizes "insertion" rather than "overlapping" to highlight the principle that all scenes are a harmonious unity, which forms a distinct transportation landscape in the city.

1) The strategy of sustaining site context

Breaking through the common design mode for transportation squares which values transportation much more than culture, the design makes full uses of human landscape resources, and highlights the cultural connotation of Nanjing. Landmarks of modern sense are created, and the city takes on a brand new look.

2) The creation of user-friendly space

To highlight the people-oriented idea, the design humanize the construction of station squares to the largest extent, which meets not only the transportation demands and but also the nearby residents' needs of living, recreation and mutual communication.

3) The sustainable landscape design

In line with the method of making full and proper use of resources, the design aims to construct a sustainable underground landscape environment with low cost and high return. The design of facilities and accessorial buildings follows the principle of low maintenance cost. A large amount of zonal plants are planted, of which deciduous trees account for a large proportion. Such design not only meets the needs of providing shades in summer and bringing in daylights in winter, but also guarantees the spatial transparency.

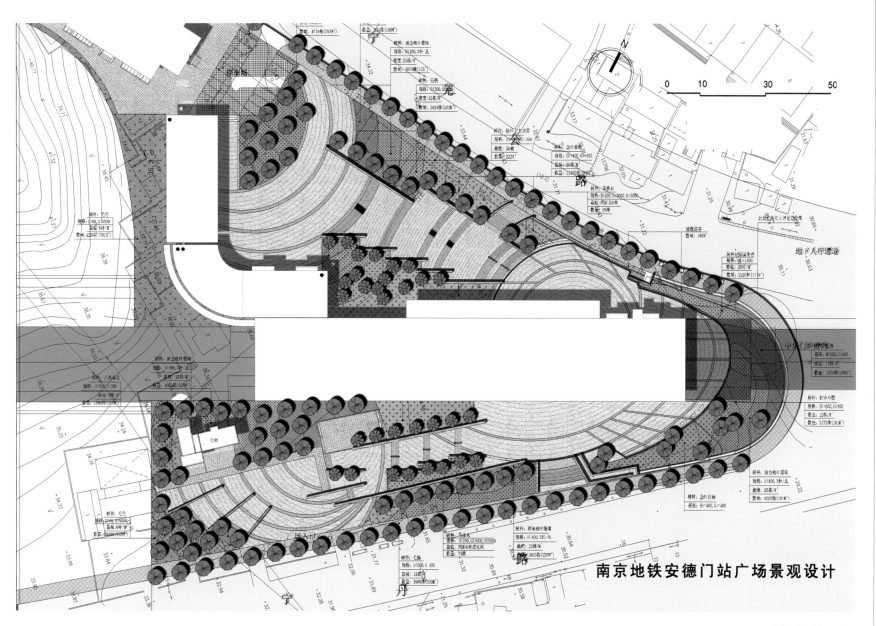

安德门地铁站总平面图

1. 安德门站实景
2. 地带性植物种植
3. 安德门站下沉广场

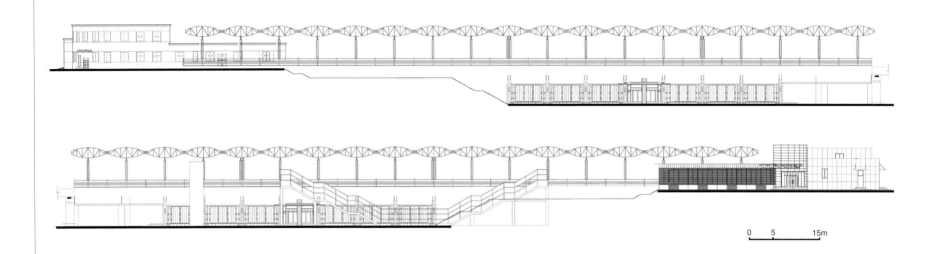

0　5　15m

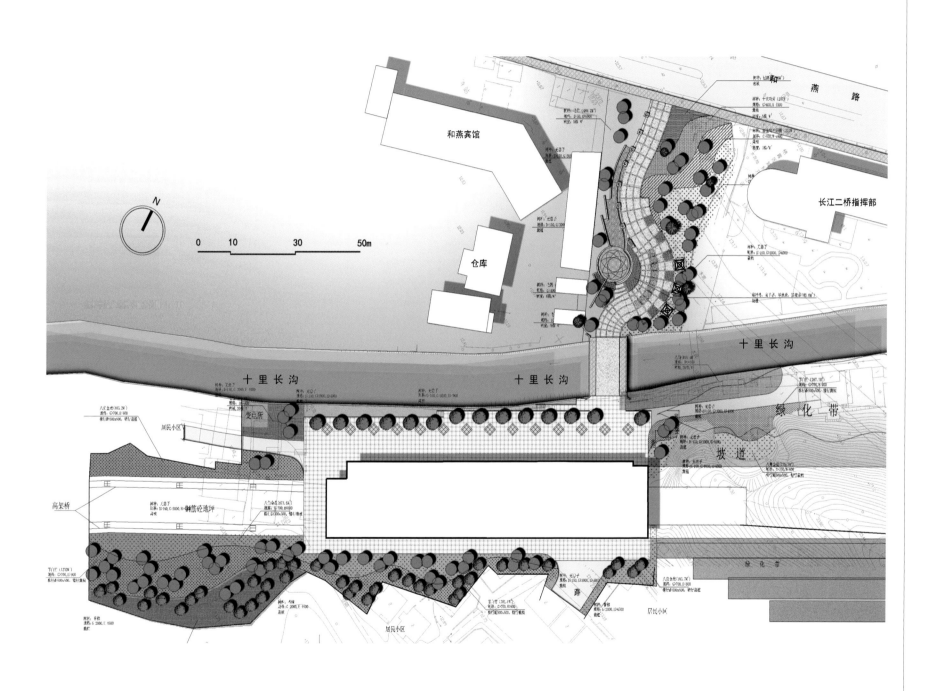

1	
3	1. 安德门站立面图
2	2. 安德门站下沉广场实景
	3. 红山动物园站总平面

1. 红山动物园站立面图
2. 红山动物园站实景
3. 红山动物园站内景

景观建筑 LANDSCAPE & ARCHITECTURE | 345

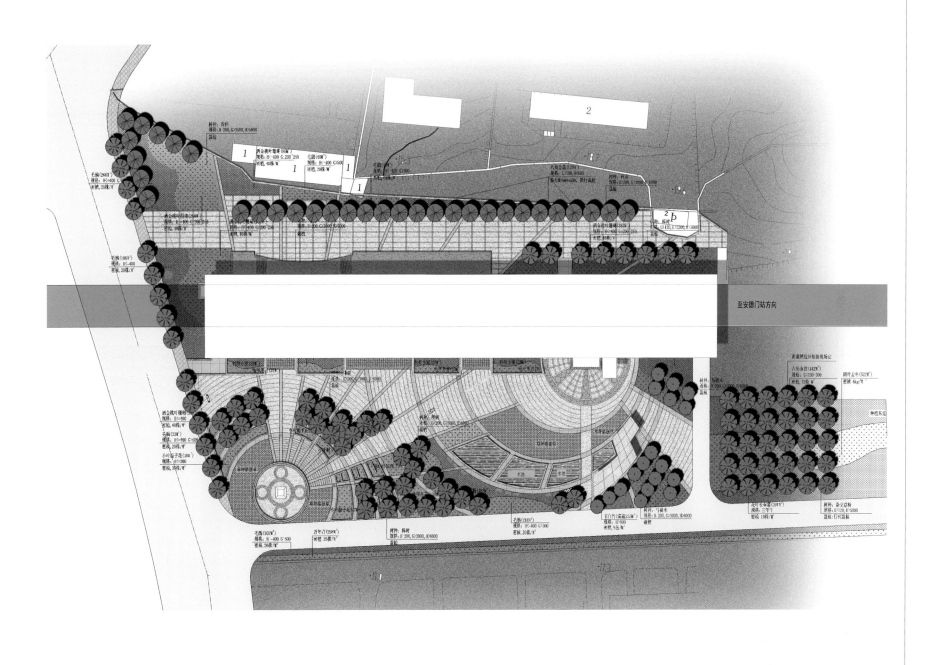

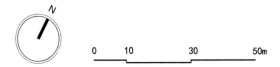

1	2		1. 红山动物园站广场休息亭
3		4	2. 红山动物园站内部实景
			3. 红山动物园站站前广场
			4. 小行站总平面

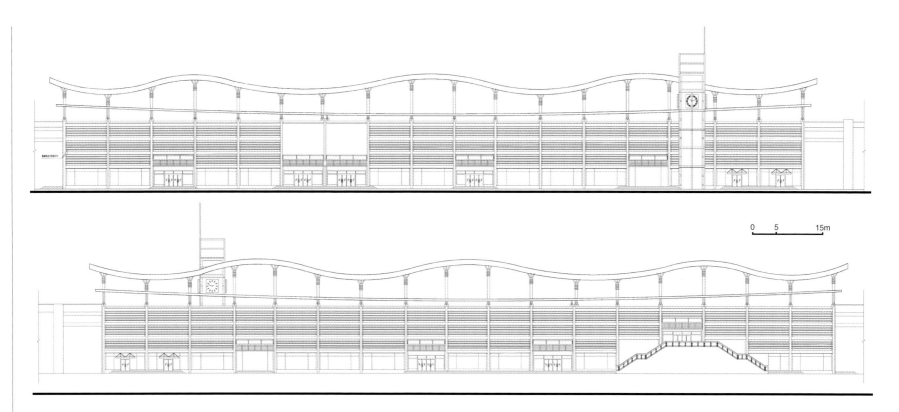

1	3	5
2	4	6

1. 小行站立面图
2. 小行站站台实景
3. 小行站建筑实景
4. 小行站前广场实景一
5. 小行站前广场实景二
6. 小行站前广场鸟瞰

1. 小行站建筑实景
2. 站前广场秋景
3. 钟塔透视

1. 迈皋桥站实景
2. 迈皋桥站站台实景
3. 迈皋桥站建筑幕墙细部

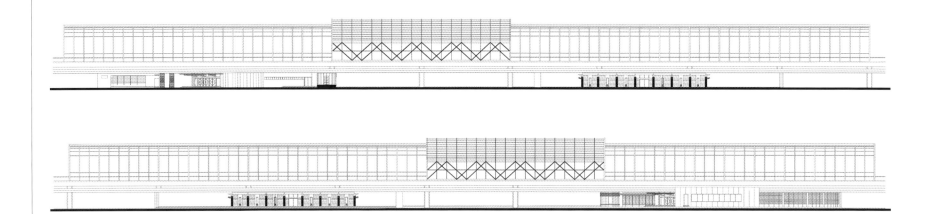

1	
	3
2	

1. 迈皋桥站立面图
2. 地铁站楼梯实景
3. 站前广场景观

0 5 15m

| 354 | LANDSCAPE & ARCHITECTURE 景观建筑

	2	1. 中华门站实景
1		2. 中华门站立面图
	3	3. 中华门站台实景

南京市紫清湖鳄鱼馆景观建筑设计
ZIQING LAKE CROCODILE EXHIBITION HALL ARCHITECTURE AND LANDSCAPE DESIGN, NANJING

项目地点：江苏省南京市	Location: Nanjing, Jiangsu	项目成员
设计时间：2010 年	Design Period: 2010	景观设计：成玉宁、袁旸洋
建成时间：2013 年	Completion Time: 2013	建筑设计：成玉宁、袁旸洋、方颖
委托单位：中宁农业科技发展公司	Client: The Agricultural Science And Technology Development Corporation	结构设计：盛春陵
用地面积：6569.4 平方米	Area: 6569.4 m²	水电设计：王晓晨、李滨海

项目概况

紫清湖生态旅游度假区位于南京市东郊汤山，总占地面积 3000 余亩，是集珍稀动物养殖、生态旅游、温泉度假于一体的度假区。鳄鱼馆位于度假区主入口处，以鳄鱼的展陈与科普为主要功能，同时提供餐饮、5D 电影等休闲服务。

设计策略

设计将建筑的主要功能与造型相结合，造型取意"远古而又神秘的生命"。

1）因势随形，表达概念

建筑设计充分契合现有地形，依势营建，造型力求简洁，建筑体块沿湖展开，富有动势。建筑屋面由南向北曲折升起，仿佛"晒太阳的鳄鱼"。

2）对话自然，丰富细部

建筑外墙采用现浇波浪状清水混凝土，粗犷与细腻、野趣与理性并存，充分对话景观环境。外立面设置连续的折线形无框长窗，强烈的虚实对比强化了建筑的体块动势与神秘感。

Project Profile

Ziqing Lake Eco-Tourism Resort is located in Nanjing's eastern suburb, Tangshang, covering a total area of over 3000 mu. The resort is a multifunctional site for rare animal cultivation, eco-tourism and hot spring experience. Crocodile Exhibition Hall stands near the main entrance, which fulfills the function of displaying crocodiles and popularizing related scientific knowledge. In addition, recreational services like catering and 5D films are also available in this hall.

Design Strategy

This deign takes a full account of the architecture's functional features. And the architectural image is derived from the concept of "ancient and mysterious life".

1) Express concepts through adjusting architectural image to the landform

In line with the existing landform, the design adjusts the construction to local conditions, with images aiming to be brief. The architecture blocks are spread along the lake, showing great kinetic potential. From south to the north, the roof coverings ride in a circuitous way, which looks like a crocodile having the sun bath.

2) Enrich details through having a dialog with the nature

The material of exterior walls is wavy bare concrete, the image of which is rough but also delicate, full of wild delight but not lacking in reasons. Such design realizes a complete dialog with the landscape environment. Moreover, frameless windows are set on the exterior walls in a continuous fold line, whose intensive comparison between virtual and true scenes reinforces the kinetic potential and mystique of the architecture blocks.

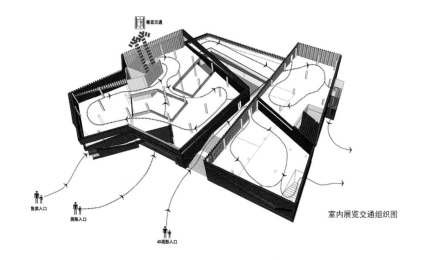

室内展览交通组织图

鳄鱼馆建筑效果图

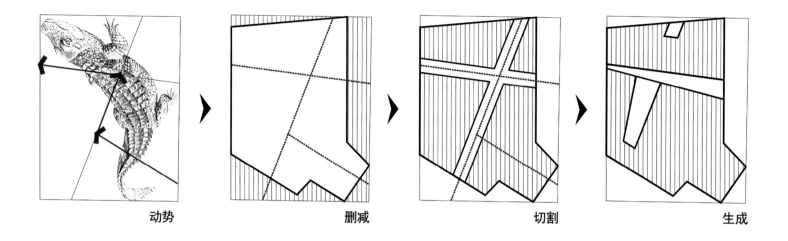

动势　　　　删减　　　　切割　　　　生成

1	2	1. 形体生成分析
		2. 体块生成分析
3		3. 建筑东侧实景

体块生成分析

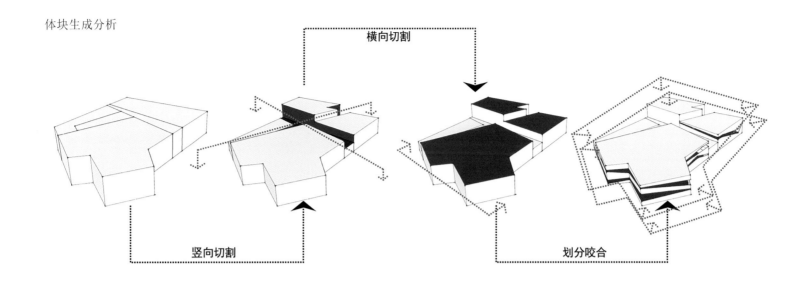

横向切割　竖向切割　划分咬合

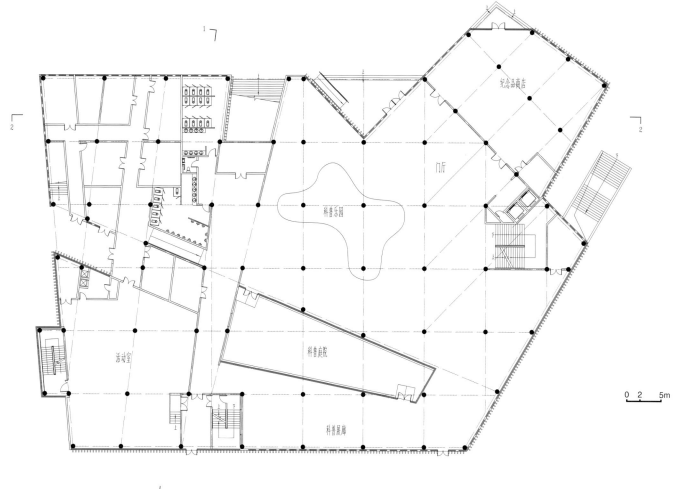
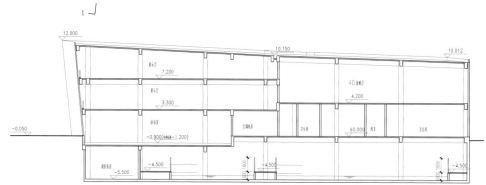
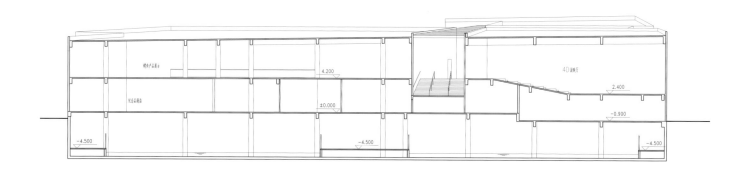

景观建筑 LANDSCAPE & ARCHITECTURE 361

1. 建筑平面图
2. 1-1 剖面图
3. 2-2 剖面图
4. 建筑立面细部

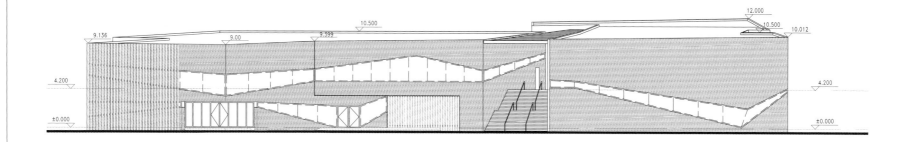
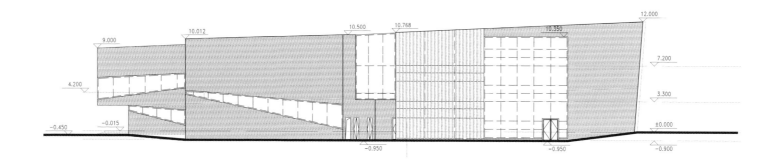
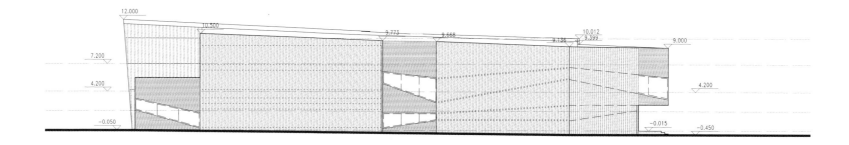
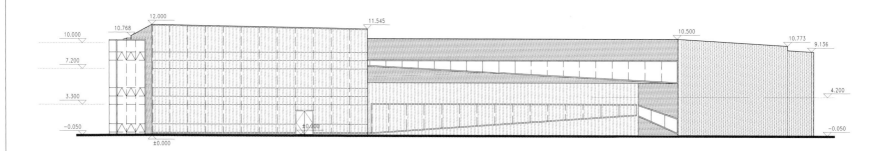

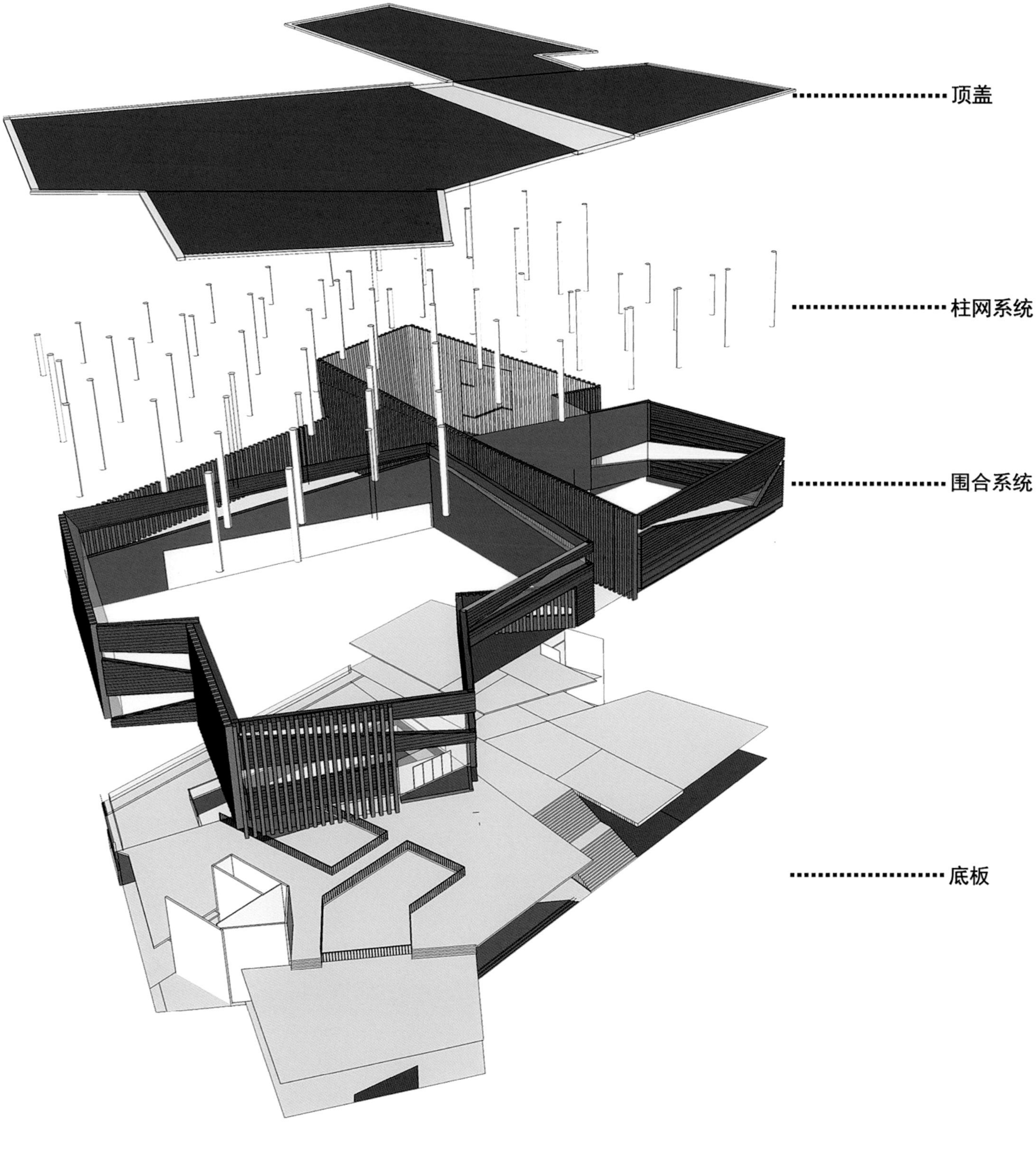

景观建筑 LANDSCAPE & ARCHITECTURE | 363

顶盖

柱网系统

围合系统

底板

1		1. 展览馆东立面
2		2. 展览馆北立面
	5	3. 展览馆南立面
3		4. 展览馆西立面
4		5. 建造结构体系

1. 展览馆西南侧实景
2. 展览馆南侧实景
3. 展览馆西侧实景

展览馆西南实景

1			1. 5D 动感影院
	3	4	2. 休闲餐厅实景
2			3. 二层走廊实景
			4. 展馆内部实景

景观建筑 LANDSCAPE & ARCHITECTURE | 371

展馆效果图

1 展馆东北侧实景

2 清水混凝土外墙细部

苏皖边区政府旧址保护规划与设计
SU-WAN BORDER REGION GOVERNMENT SITE CONSERVATION PLANNING AND DESIGN

项 目 地 点：江苏省淮安市	Location: Huai'an, Jiangsu	项 目 成 员
设 计 时 间：2010年	Design Period: 2010	景 观 设 计：成玉宁、李雯婷
建 成 时 间：2012年	Completion Time: 2012	建 筑 设 计：成玉宁、李雯婷
委 托 单 位：淮安市园林局	Client: Jiangsu, Huai'an Garden Bureau	结 构 设 计：薛 峰
用 地 面 积：72158平方米	Area: 72158 m²	水 电 设 计：刘 俊、钱荣银

项目概况

地处淮安的"苏皖边区政府"在新民主主义革命时期具有重要的历史地位，曾获得毛泽东主席的高度评价，现为全国重点文物保护单位。旧址仅有原"苏皖边区政府交际处"保存较为完好，设计结合既有建筑与遗存保护，增建了史料陈列馆。

设计策略

设计的重点在于把握历史建筑保护与更新的关系，以处理有形的物质形态与场所精神的关系、单体与整体的关系以及整体氛围的复原与特定时代精神的展现这三个方面为聚焦点。

满足新、老建筑的"对话"，以旧址的"大宅"为原型，"复活"了具有典型意义和历史价值的环境。以"融入"作为设计手段，西向与历史建筑相融合、东向兼顾城市空间、南北向与历史街区相协调。新馆的比例、质感、色彩等力求与历史遗存相呼应。建筑风格秉承淮安地方特色，采取院落格局，以二层建筑为主，局部出挑，高低错落、变化有致。展陈建筑设东、西两路轴线，东路的单体建筑围绕庭院呈行列式布置；西路院落空间变化丰富。为满足展陈需要，适当放大了陈列馆建筑的尺度。展厅结合院落逐次展开，各展陈空间相对独立又彼此呼应。通过打通新、老建筑交际处的围墙并增建复廊，以沟通东西。"传神"与"细节"的表达，使新、老建筑成了一个有机的整体。

Project Profile

Located in Huai'an, "Su-Wan Border Region Government" used to earn a high appraisal from Chairman Mao for its value in the new-democratic revolution. Now, it has become China's national relic protection unit. Since only the border region government site is well-preserved, this design blends construction with relic protection, and builds out a history museum.

Design Strategy

The design makes great efforts in dealing with the relationship between the protection and renewal of the historical architectures, with a focus on the relationship between physical form and site spirits, the relationship between individual and unity, the renovation of the overall atmosphere and the manifestation of spirit for a specific era.

On the basis of conducting dialogs between old and new architectures, the design uses the former site of "Dazhai" as the prototype to "revitalize" the environment which contains typical meanings and historical values. With "integration" as the design method, the west section is blended with historical architectures, the east section takes city space into account, and the north-south part is in coordination with the historical streets. The new museum is designed to be in harmony with historic relics in all aspects, such as scale, ratio, texture and color. The architectural style bears regional features of Huai'an, which adopts courtyard pattern and is always dominated by two-story buildings. With some parts highlighted, an uneven and changeable image is created. The exhibition hall is equipped with two axes respectively in the east and the west. Along the east axis, individual buildings are arranged around courtyards in rows. And the courtyard space along the west axis shows great varieties. To meet the demands of exhibition, the architecture scale is enlarged to some extent. The exhibition halls sequentially spread with the courtyards, which makes each hall both independent and interconnected. The design connects the east and the west through tearing down the enclosing wall at the border between the old and new architectures. The "vivid" and "detailed" expression integrates the old and new architectures into an organic unity.

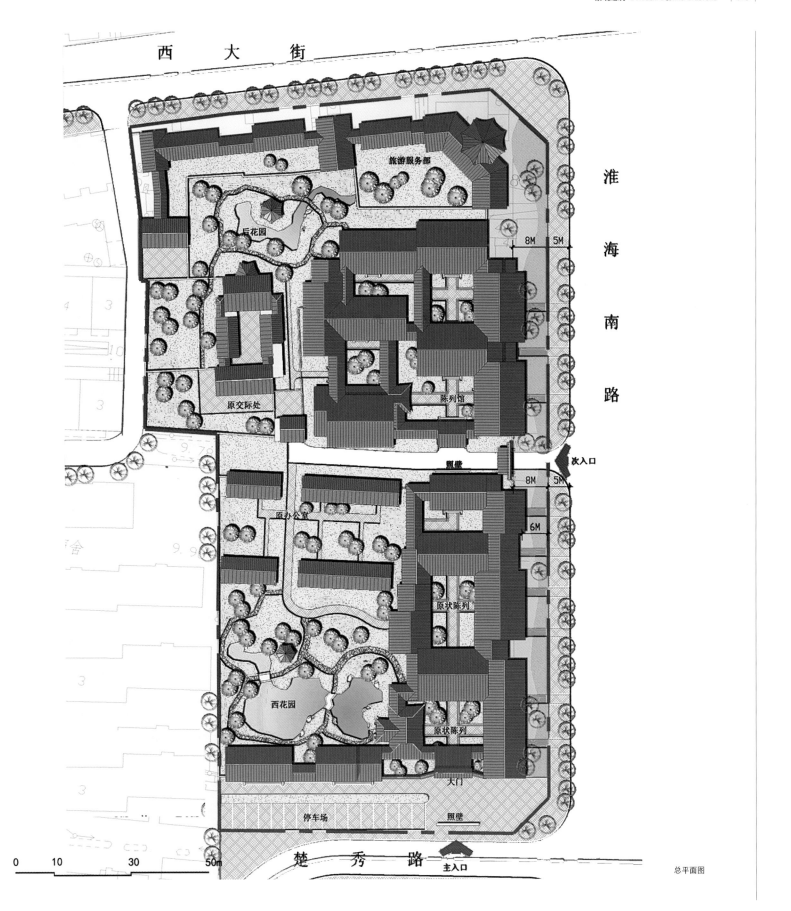

总平面图

1		1. 建筑主入口实景
2	3	2. 领导人集体群雕
		3. 庭院实景

| 1 | 2 | 1. 前院实景
2. 后院实景 |

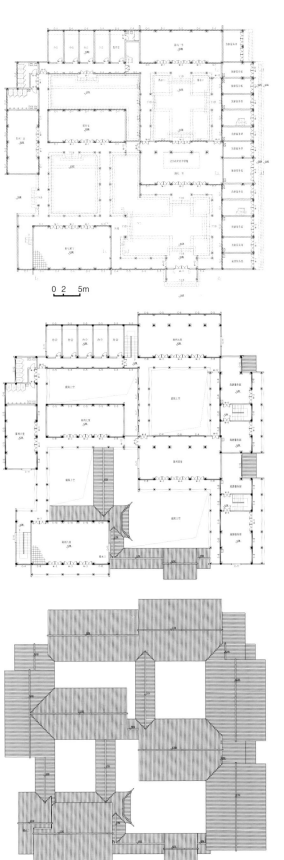

1. 前院实景
2. 一层平面
3. 二层平面
4. 屋顶平面

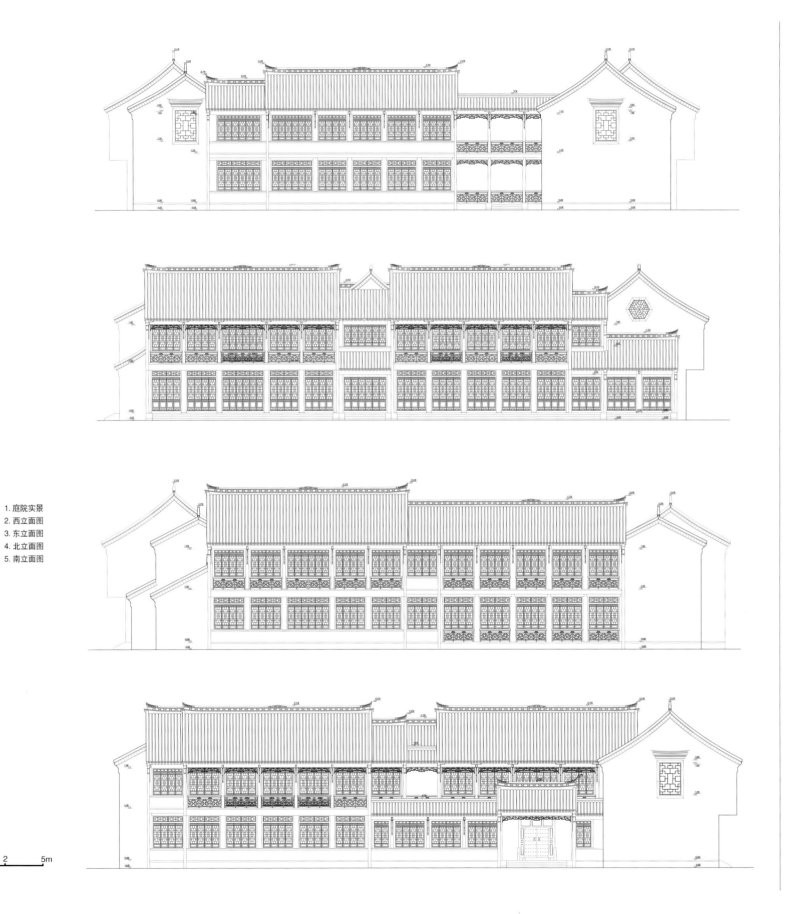

1. 庭院实景
2. 西立面图
3. 东立面图
4. 北立面图
5. 南立面图

1. 西院实景一
2. 西院实景二
3. 二层回廊实景

常熟市古里镇坞丘农耕文化馆
WUQIU FARMING CULTURE CENTER OF GULI TOWN, CHANGSHU

项目地点：常熟市古里镇	Location: Guli Town, Changshu	项目成员
设计时间：2010年	Design Period: 2010	景观设计：成玉宁、许立南
建成时间：2011年	Completion Time: 2011	建筑设计：成玉宁、许立南、张李瑞
委托单位：常熟市古里镇人民政府	Client: People's Government of Guli Town, Changshu	结构设计：王伟成
用地面积：用地7458.4平方米，建筑面积为2480平方米	Area: Land area 7,458.4 ㎡, Construction area 2,480 ㎡	水电设计：王晓晨、李滨海

项目概况

坞丘农耕文化馆位于常熟市古里镇的农田之中，总用地面积7458.4m²。设计在协调场地整体空间的前提下，凸显场所生态特色、彰显农业文明，营造了具有地方特色的现代农机具展示馆舍。

设计策略

农耕文化馆主体为一层建筑，局部两层，集农机仓库、产品展销、办公等多种功能于一体。设计结合场所条件，以常熟农业文化为背景，提取当地传统建筑作为基本元素，植入新建筑之中。建筑风格质朴而富有乡土气息，整体形态呈半围合之势，充分对话田野。

Project Profile

Wuqiu Farming Culture Center is located in the farmlands of Guli Town, Changshu, covering an area of 7458.4m². With coordinating the site's overall space as the precondition, the design highlights the site's ecological features and farming civilization, constructing exhibition buildings with regional characteristics for modern agricultural implements.

Design Strategy

With most buildings one-story, some two-story, Farming Culture Center can function as agricultural implements warehouse, product exhibition hall, and office building. With Changshu's agricultural culture as the background, the design utilizes the site conditions, and extracts local traditional architectural factors to inject into new buildings. In this way, the architectural style is plain and full of local flavors. The overall architectural shape is half enclosed so as to have close connection with the farmlands.

设计构思草图

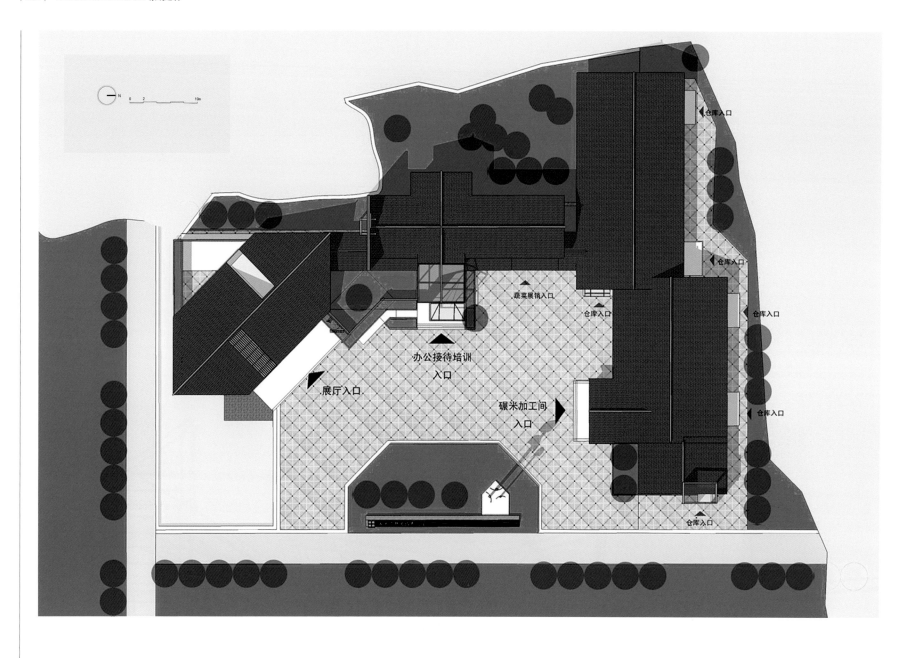

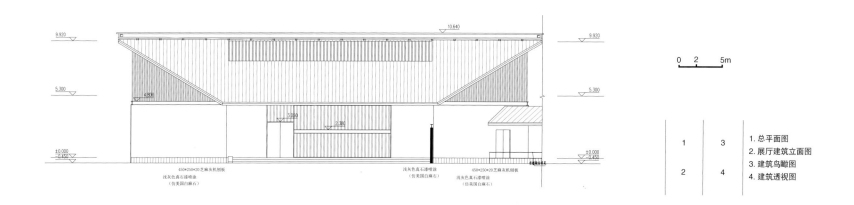

1. 总平面图
2. 展厅建筑立面图
3. 建筑鸟瞰图
4. 建筑透视图

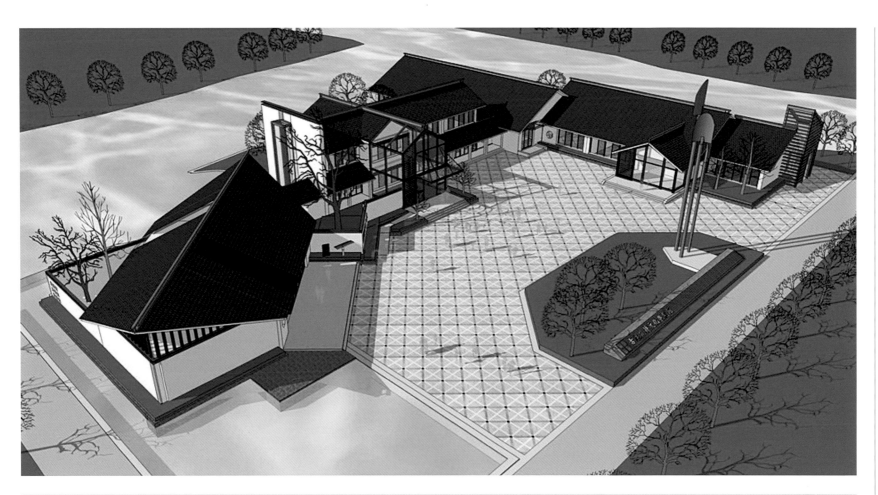

农耕文化馆鸟瞰效果图

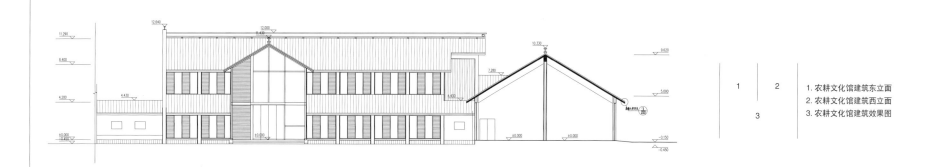

1	2	1. 农耕文化馆建筑东立面
		2. 农耕文化馆建筑西立面
3		3. 农耕文化馆建筑效果图

394 LANDSCAPE & ARCHITECTURE 景观建筑

农耕文化馆实景

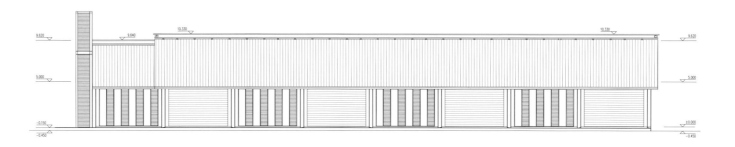

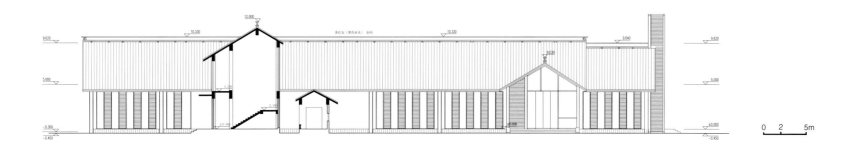

1. 农耕文化馆北立面
2. 农耕文化馆剖立面
3. 农耕文化馆实景

淮安市龙光阁设计
LONGGUANG PAVILION DESIGN, HUAI'AN

项 目 地 点：江苏省淮安市	Location: Huai'an , Jiangsu	项 目 成 员
设 计 时 间：2008 年	Design Period: 2008	景 观 设 计：成玉宁、张 楷
建 成 时 间：2014 年	Completion Time: 2014	建 筑 设 计：成玉宁、张 楷
委 托 单 位：淮安市规划局楚州分局	Client: Huai'an planning bureau Chuzhou branch	结 构 设 计：王伟成
用 地 面 积：约 30100 平方米	Area: About 30100 ㎡	水 电 设 计：王晓晨、李滨海

项目概况

古龙光阁位于淮安城外东南方向的护城冈上，原建筑由明末漕督朱大典所建，于清道光二十四年重修，在抗日战争时期毁于战火，现址为古城墙遗址公园。龙光阁的重建不仅再现了淮安名胜，对于激活城市地块、推动老城的旅游发展有着重要意义。

设计策略

设计以历史上的龙光阁为原型，营造了具有典型意义和历史价值的地标建筑。设计着力表现淮安的传统建筑特色，采取庭院与楼阁相结合的方式，阁下因山就势设有辅房，以爬山廊连接。建筑主体采用钢筋混凝土结构，门窗等小木作均为实木。

在满足新建筑的功能需求的基础上，设计重点协调了建筑与周边环境之间的对话关系：向南与淮安入城道路形成对景，向西与古城墙呼应，适当改造原有的地形、水系，使建筑更好地融于周边环境。此外，景观设计结合用场地条件及周边状况，合理布置功能区，突出历史遗址的保护与展示功能，彰显场所特征。

Project Profile

The old Longguang Pavilion is located on the defense hillock to the southeast of Huai'an city. First built by Zhu Dadian, councilor for water transportation in the late Ming dynasty, the original pavilion was renovated in the 24th year since Emperor Daoguang ruled the Qing dynasty. Unfortunately, it was ruined by the flames of the Anti-Japanese war. The current site is a heritage park of the ancient city wall. The reconstruction of Longguang Pavilion not only represents the famous scenic spot of Huai'an, but also has great importance to revitalizing the city blocks and promoting the old city's tourism development.

Design Strategy

With the original Longguang Pavilion as the prototype, this design constructs a landmark architecture which carries typical connotation and historical values. The construction emphasis is put on the dialog relationship between the architecture and its surrounding environment. The reconstructed pavilion forms an opposite scene with the city entrance road in the south, and adds radiance and beauty to the ancient city wall in the west. The landform and water system are properly remolded to better assimilate the architecture into the surrounding environment. In addition, the landscape design utilizes the site conditions and its surrounding states to make appropriate arrangement of function zones, which highlights the protection and exhibition of the historic relics as well as the site features.

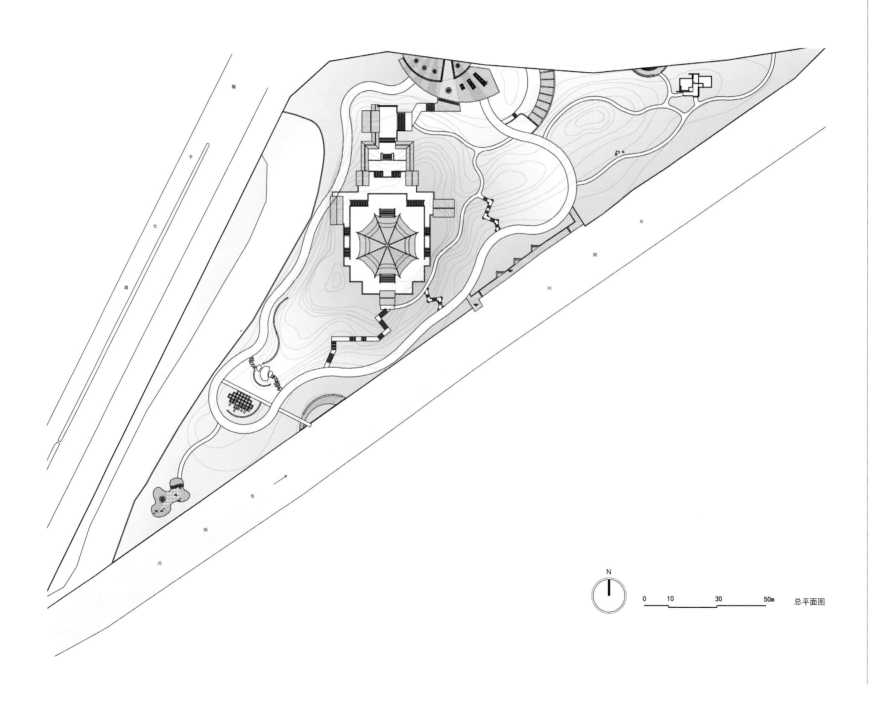

总平面图

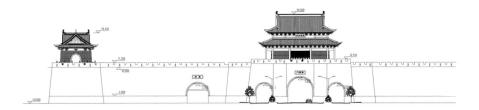
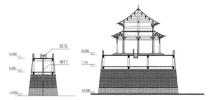

1. 城垣博物馆二层平面图
2. 城垣博物馆一层平面图
3. 城垣立面、剖面图
4. 城门、瓮关实景
5. 龙光阁实景

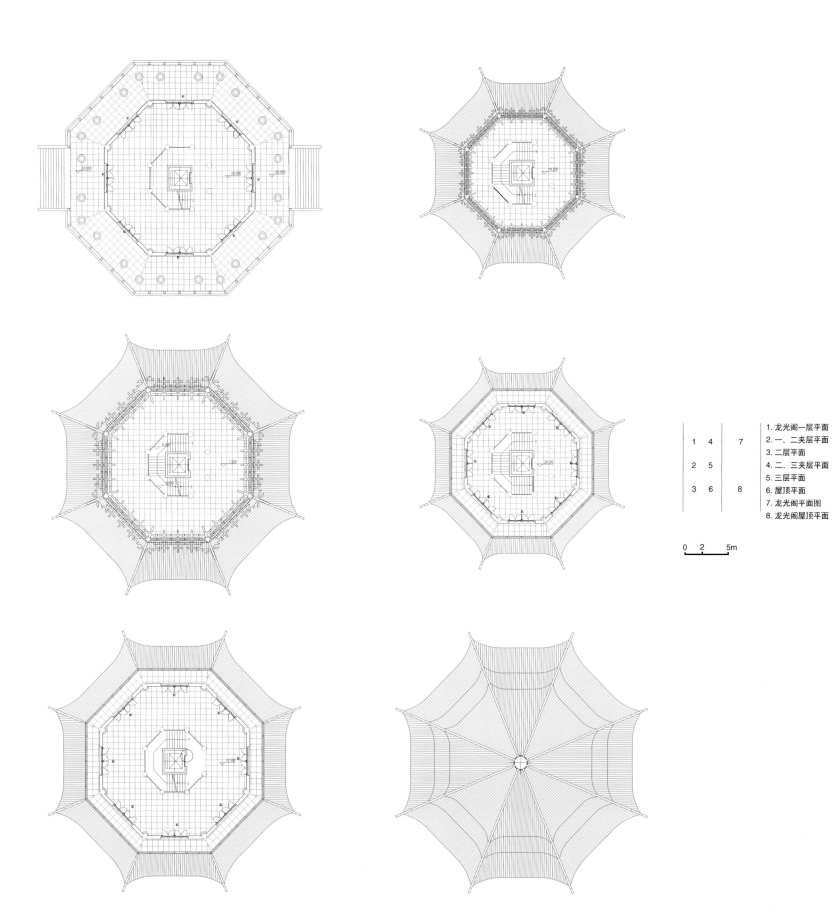

1. 龙光阁一层平面
2. 一、二夹层平面
3. 二层平面
4. 二、三夹层平面
5. 三层平面
6. 屋顶平面
7. 龙光阁平面图
8. 龙光阁屋顶平面

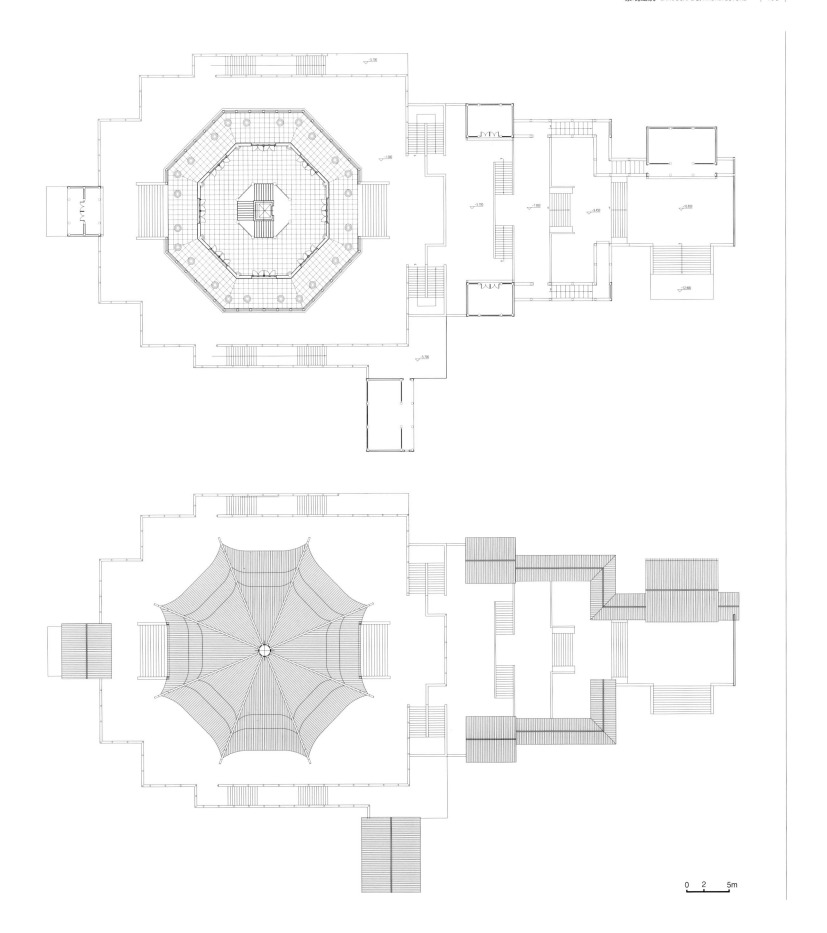

龙光阁实景

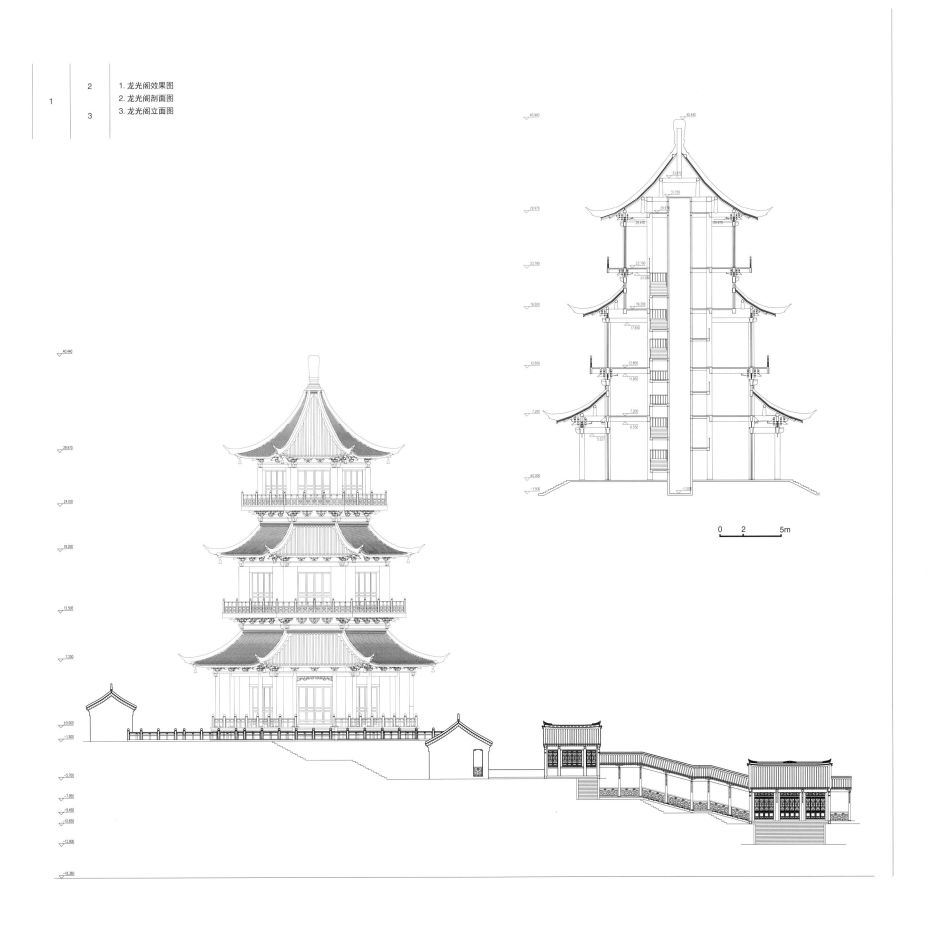

1. 龙光阁效果图
2. 龙光阁剖面图
3. 龙光阁立面图

后记

光阴荏苒，过去的30年是中国改革开放取得突出成就的重要时期，社会经济取得举世瞩目的成就，华夏民族在不断创造人类的奇迹。伴随着人居环境的建设，当代中国风景园林事业获得了空前地发展，与此同时也引发了新的思索：现代化、民族性、国际化、地域主义等始终是过去30年中国风景园林界思考的命题。

30年之于中国风景园林的历史是短暂的瞬间，但对于个人而言却是职业生涯的重要时期，很难以时间的长短加以度量。从摸索前行到生成设计理念、方法、体系，这一切都是在特定的时空维度中展开的，因此时间具有了特殊的意义。35年前笔者有幸接触了风景园林并接受了专业的学习，本科四年的学习伴随着"兴奋与困惑"度过，思维的"自由"与设计的"依据"一如跷跷板，巧妙地把握动态平衡点成为初学者的门槛；30年前开始有可能是根据自己对于项目与场地的理解及意愿而展开设计，随之而来的是面临着："图纸与建造""方案与方法""科学与艺术""传承与创新"等问题；而近20年来笔者似乎更加关注："特征与场所""国际与地域""定性与定量""尺度与逻辑""耦合与拼贴""本体与形式""形而下与形而上"等一系列问题。不同时间段关注的焦点在发生着变化，由表及里、由易到难，符合认知的逻辑。所有这一系列问题似乎都以成对的范畴出现，涉及景园观、认知论、方法论等层面，我们面临的问题不是简单的取舍，而是透过权衡利弊以寻求设计的智慧。沿着实践的线索，问题与答案都是在实践与思考中得以解决与印证。有思考的设计与基于设计的思考是解决问题的根本所在，坚持两条腿走路、走中国特色的风景园林之路是笔者30年不渝的信念。

将过去的实践与设计理念集结成书奉献给读者，藉此也为今后个人的设计之路进一步廓清方向。这里笔者想呈现的绝不仅仅是个人的实践，更希望探索的是具有这个时代印记的中国新园林，不论成就高下，付出无悔。

在此感谢中国工程院孟兆帧院士的厚爱并惠为本书作序；感谢东南大学潘谷西、杜顺宝及培养我的导师们；感谢中国建筑工业出版社陈桦编审及杨琪的支持；感谢江苏省住建厅张晓鸣处长以及赖自力、陈家松先生协助部分项目摄影；感谢宾夕法尼亚州立大学Ron Henderson教授、华东师范大学象伟宁教授审阅本书目录及序的英文译稿；感谢相关业主及施工单位的朋友们；感谢弟子袁旸洋、谭明、单梦婷、鲍洁敏为本书出版付出的心力；感谢家人的长期理解与支持。

于东南大学逸夫建筑馆303
2015年9月9日

POSTSCRIPT

Time flies. The past 30 years is an important period when China has obtained outstanding achievements through reform and opening-up to the outside world. While the whole world's attention is attracted by China's social and economic accomplishments, the Chinese nation never stops its step in creating miracles for human beings. With the construction of human settlement, unparalleled development has been achieved in the cause of contemporary Chinese landscape architecture. Meanwhile, the past 30 years also brings new reflections on modernization, national characters, internalization, and regionalism, which have been the major subjects studied by the world of Chinese landscape architecture.

Compared with the history of Chinese landscape architecture, 30 years is just a moment. But as for individuals, 30 years is an important period for their career development which cannot be simply measured by length. From groping in the dark to generating design philosophies, methods and system, everything is unfolded in a particular spatial dimension, thus time is endowed with special meanings. 35 years ago, I was fortunate to get in touch with landscape architecture and received professional education. The four-year undergraduate study was finished with "excitement and confusion". The "freedom" of thinking and the "foundation" of design seemed to sit at each end of a seesaw, so grasping the dynamic equilibrium point became the threshold in front of beginners. At the beginning 30 years ago, my design was based on my personal interpretation of the project and the site, followed by problems, such as the contradiction between "drawing and construction", "plan and method", "science and art", "inheritance and innovation". During the recent 20 years, I pay more attention to a series matters like "features and the site", "the globe and regions", "quality and quantity", "scale and logistics", "interconnection and patching", "entity and form", "metaphysics and physics". The focus has changed at different stages, from superficial to fundamental, from easy to difficult, which conforms to the logistics of cognition. All appearing in pairs, these matters involve many aspects of landscape architecture outlook, cognitive theory and mythology. What we need to do is not simply accepting or rejecting, but seeking the wisdom of design after weighing the merits and demerits. Only through practicing and thinking can problems be solved and answers be verified. Design with thinking and thinking based on design are fundamental for solving problems. It's my long-standing faith for 30 years to unswervingly take the road of landscape architecture with Chinese characteristics.

This book aims to show readers the fruit of my former practices and design ideas, and also to make clear the further direction of my design. What I want is by no means only presenting my personal practices, but exploring new Chinese landscape marked with features of this era. I will never regret for my efforts made in this filed whether I can get any return.

Hereby, I would like to extend my sincere gratitude to Mr. Meng Zhaozhen, academician of the Chinese Academy of Engineering, for his great kindness to this book and efforts in writing the preface; to Ms. Chen Hua, copy editor of China Achitecture & Building Press, and her accociate Ms. Yang Qi, for their support; to Mr. Zhang Xiaoming, section chief of Department of Housing and Urban-Rural Development of Jiangsu Province, and Mr. Lai Zili and Mr. Chen Jiasong, for their help in photographing some projects; to Prof. Ron Henderson from University of Pennsylvania , Pro. Xiang Weining from East China Normal University, for their efforts in reviewing the English translation of the catalogue and the preface; to related proprietors and friends from units in charge of the construction; to my students, Yuan Yangyang, Tan Ming, Shan Mengting, and Bao Jiemin, for their contributions for publishing this book; and to my family, for their long-term understanding and support.

In room 303 of Yifu Architecture Building, Southeast University
September 9, 2015